T0205402

Lecture Notes in Networks and Systems 683

Series Editor

Janusz Kacprzyk, *Systems Research Institute, Polish Academy of Sciences, Warsaw, Poland*

Advisory Editors

Fernando Gomide, *Department of Computer Engineering and Automation—DCA, School of Electrical and Computer Engineering—FEEC, University of Campinas—UNICAMP, São Paulo, Brazil*
Okyay Kaynak, *Department of Electrical and Electronic Engineering, Bogazici University, Istanbul, Türkiye*
Derong Liu, *Department of Electrical and Computer Engineering, University of Illinois at Chicago, Chicago, USA*
Institute of Automation, Chinese Academy of Sciences, Beijing, China
Witold Pedrycz, *Department of Electrical and Computer Engineering, University of Alberta, Alberta, Canada*
Systems Research Institute, Polish Academy of Sciences, Warsaw, Poland
Marios M. Polycarpou, *Department of Electrical and Computer Engineering, KIOS Research Center for Intelligent Systems and Networks, University of Cyprus, Nicosia, Cyprus*
Imre J. Rudas, *Óbuda University, Budapest, Hungary*
Jun Wang, *Department of Computer Science, City University of Hong Kong, Kowloon, Hong Kong*

The series "Lecture Notes in Networks and Systems" publishes the latest developments in Networks and Systems—quickly, informally and with high quality. Original research reported in proceedings and post-proceedings represents the core of LNNS.

Volumes published in LNNS embrace all aspects and subfields of, as well as new challenges in, Networks and Systems.

The series contains proceedings and edited volumes in systems and networks, spanning the areas of Cyber-Physical Systems, Autonomous Systems, Sensor Networks, Control Systems, Energy Systems, Automotive Systems, Biological Systems, Vehicular Networking and Connected Vehicles, Aerospace Systems, Automation, Manufacturing, Smart Grids, Nonlinear Systems, Power Systems, Robotics, Social Systems, Economic Systems and other. Of particular value to both the contributors and the readership are the short publication timeframe and the world-wide distribution and exposure which enable both a wide and rapid dissemination of research output.

The series covers the theory, applications, and perspectives on the state of the art and future developments relevant to systems and networks, decision making, control, complex processes and related areas, as embedded in the fields of interdisciplinary and applied sciences, engineering, computer science, physics, economics, social, and life sciences, as well as the paradigms and methodologies behind them.

Indexed by SCOPUS, INSPEC, WTI Frankfurt eG, zbMATH, SCImago.

All books published in the series are submitted for consideration in Web of Science.

For proposals from Asia please contact Aninda Bose (aninda.bose@springer.com).

Arthur Gibadullin
Editor

Digital and Information Technologies in Economics and Management

Proceedings of the International Scientific and Practical Conference "Digital and Information Technologies in Economics and Management" (DITEM2022)

Editor
Arthur Gibadullin
Moscow Power Engineering Institute
National Research University
Moscow, Russia

ISSN 2367-3370 ISSN 2367-3389 (electronic)
Lecture Notes in Networks and Systems
ISBN 978-3-031-30925-0 ISBN 978-3-031-30926-7 (eBook)
https://doi.org/10.1007/978-3-031-30926-7

© The Editor(s) (if applicable) and The Author(s), under exclusive license
to Springer Nature Switzerland AG 2023
This work is subject to copyright. All rights are solely and exclusively licensed by the Publisher, whether the whole or part of the material is concerned, specifically the rights of translation, reprinting, reuse of illustrations, recitation, broadcasting, reproduction on microfilms or in any other physical way, and transmission or information storage and retrieval, electronic adaptation, computer software, or by similar or dissimilar methodology now known or hereafter developed.
The use of general descriptive names, registered names, trademarks, service marks, etc. in this publication does not imply, even in the absence of a specific statement, that such names are exempt from the relevant protective laws and regulations and therefore free for general use.
The publisher, the authors, and the editors are safe to assume that the advice and information in this book are believed to be true and accurate at the date of publication. Neither the publisher nor the authors or the editors give a warranty, expressed or implied, with respect to the material contained herein or for any errors or omissions that may have been made. The publisher remains neutral with regard to jurisdictional claims in published maps and institutional affiliations.

This Springer imprint is published by the registered company Springer Nature Switzerland AG
The registered company address is: Gewerbestrasse 11, 6330 Cham, Switzerland

Preface

The conference was held with the aim of summarizing international experience in the field of digital development of the economy and management, the introduction of information technologies, and systems in the organizational processes of managing corporations and individual industries.

The II International Scientific and Practical Conference "Digital and Information Technologies in Economics and Management" (DITEM2022) was held on November 21–23, 2022.

The conference addressed issues of information and digital and intellectual technologies in economics and management. A distinctive feature of the conference is that it presented reports of authors from China, Saudi Arabia, Tunisia, Uzbekistan, Tajikistan, Angola, Kazakhstan, India, and Russia. Researchers from different countries presented the process of transition of economic activities to the information and digital path of development and presented the main directions and developments that can improve the efficiency and development of the economy and management.

The conference sessions were moderated by Gibadullin Arthur of the National Research University "Moscow Power Engineering Institute", Moscow, Russia.

Thus, the conference still facilitated scientific recommendations on the use of information, computer, digital and intellectual technologies in industry, and fields of activity that can be useful to state and regional authorities, international and supranational organizations, and the scientific and professional community.

Each presented paper was reviewed by at least three members of Program Committee in a double-blind manner. As a result of the work of all reviewers, 16 papers were accepted for publication out of the 42 received submissions. The reviewers were based on the assessment of the topic of the submitted materials, the relevance of the study, the scientific significance and novelty, the quality of the materials, and the originality of the work. Reviewers, Program Committee members, and Organizing Committee members did not enter into discussions with the authors of the articles.

Each presented paper was reviewed by at least three members of Program Committee in a double-blind manner. As a result of the work of all reviewers, 16 papers were accepted for publication out of the 41 received submissions. The reviews were based on the assessment of the topic of the submitted materials, the relevance of the study, the scientific significance and novelty, the quality of the materials, and the originality of the work. Reviewers, Program Committee members, and Organizing Committee members did not enter into discussions with the authors of the articles.

The Organizing Committee of the conference expresses its gratitude to the staff at Springer who supported the publication of this proceedings. In addition, the Organizing Committee would like to thank the conference participants, reviewers, and everyone who helped organize this conference and shape the present volume for publication in the Springer LNNS series.

Arthur Gibadullin

Organization

Program Committee Chairs

Asrorzoda Ubaydullo	International University of Tourism and Entrepreneurship of Tajikistan, Tajikistan
Sadriddinov Manuchehr	Branch of the International University of Tourism and Entrepreneurship of Tajikistan in the Sughd Region, Tajikistan
Jabborov Bahriddin	International University of Tourism and Entrepreneurship of Tajikistan, Tajikistan
Gibadullin Arthur	National Research University "Moscow Power Engineering Institute", Russia

Program Committee

Shmanev Sergey	European Academy of Natural Sciences, Germany
Łakomiak Aleksandra	Wroclaw University of Economics and Business, Poland
Bazarov Orifzhan Shadievich	Karshi Engineering and Economic Institute, Uzbekistan
Tarun Chakravorty N. N.	Canadian University of Bangladesh (CUB), Bangladesh
Firsov Yury	Prague Institute for Advanced Studies, Czech Republic
Liu Zi Feng	China University of Petroleum-Beijing at Kramay, China
Uzakov Gulom Norboevich	Karshi Engineering and Economic Institute, Uzbekistan
Gibadullin Arthur	National Research University "Moscow Power Engineering Institute", Russia
Rabizoda Najibullo	International University of Tourism and Entrepreneurship of Tajikistan, Republic of Tajikistan
Morkovkin Dmitry	Financial University under the Government of the Russian Federation, Russia
Davlatov Davlatmakhmad	Mining and Metallurgical Institute of Tajikistan, Tajikistan

Fakhrutdinov Irek	Academy of Sciences of the Republic of Tatarstan, Tatarstan
Pulyaeva Valentina	Financial University under the Government of the Russian Federation, Russia

Organizing Committee

Sadriddinov Manuchehr	Branch of the International University of Tourism and Entrepreneurship of Tajikistan in the Sughd Region, Tajikistan
Gibadullin Arthur	National Research University “Moscow Power Engineering Institute”, Russia
Morkovkin Dmitry	Financial University under the Government of the Russian Federation, Russia
Pulyaeva Valentina	Financial University under the Government of the Russian Federation, Russia

Organizer

International University of Tourism and Entrepreneurship of Tajikistan, Tajikistan

Contents

Numerical Integration and Numerical Solution of Differential Equations in the MatLab Digital Computing Environment

Dmitry A. Kurasov(✉) and Egor K. Karpov

University of Tyumen, St. Volodarskogo, 6, Tyumen 625003, Russia
naukka@mail.ru

Abstract. The article presents questions related to the difficulty of numerical integration and the difficulty of numerical solution of differential equations, considers the simplest ordinary differential equations and quadrature formulas, implements a software complex on a computer in the MatLab software complex, obtained indicators of calculation of a certain given function on this spectrum, as well as its clear direct solution. The values calculated in different ways were compared, as well as absolute errors associated with the number of partitions of the integration segment during numerical integration.

Keywords: Numerical Differentiation · Numerical Integration · MatLab · Ordinary Differential Equation · Trapezoid Method · Rectangle Method

1 Introduction

Simulation modeling is a more global and comprehensive way to analyze and evaluate the performance of systems whose behavior depends on the influence of unforeseen causes. The implementation of the simulation process is virtually impossible without the introduction of computer equipment. Thus, it does not matter what kind of simulation model has. Ultimately, it will be more or less complex as a digital software product.

MatLab, abbreviated from English "Matrix Laboratory", is a package of application programs for solving technical problems, as well as the programming language of the same name used for this package. The software complex MatLab [1, 2] in terms of functionality can be attributed to a mid-range product capable of solving problems using a formalized mathematical language. On the other hand, the complex can solve a wide range of problems in the field of engineering analysis, the so-called CAE systems. This indicates the versatility of this product. In the environment of a higher educational institution, it is a standard tool or module [3] for work in the field of mathematics, engineering and science. In industry, MatLab is a tool for high-performance research and development as well as data analysis [4–8]. Historically, the MatLab system, as a system for automated calculation of mathematical dependencies, appeared one of the first. This circumstance characterizes the product as one of the most reliable. It is based on the programming language with syntax based on extended matrix operations. Matrices have

© The Author(s), under exclusive license to Springer Nature Switzerland AG 2023
A. Gibadullin (Ed.): DITEM 2022, LNNS 683, pp. 1–14, 2023.
https://doi.org/10.1007/978-3-031-30926-7_1

extensive applications in complex algorithmic problems that require the use of mathematical dependencies of increased complexity. Examples are linear algebra problems and problems of creating mathematical models that have a static and dynamic behavior of system parameters and system elements. The solution of such problems boils down to the automatic complex compilation of equations of a system of parameters corresponding to a dynamic object or system as a whole. Today's development of the MatLab program suggests that the software complex has gone beyond matrix specialization and is a powerful integrated universal calculation system. The term "integrated" or "embedded" corresponds to the presence of a universal and user-friendly shell, the presence of an expression and text explanation editor, the presence of a solver and a graphical software microprocessor.

The necessary advantages of the system are its transparency and extensibility. With all this, the personality of the MatLab package will be that the user needs to know about it exactly as much as the task being solved asks. For example, in a simple case, MatLab can play the role of a common tool for performing the simplest mathematical operations, for which the symbolism of mathematical operations is sufficient. If solving the problem requires some special tools, then MatLab offers the user an almost multifunctional object-oriented programming language combined with interactive debugging tools for designed programs.

It should be noted that the MatLab software complex has a Simulink [9] calculation environment for modeling and design using models corresponding to embedded and dynamic systems. It implements the principle of visual programming, that is, without the use of program code, it allows you to synthesize the logic of the control system of increased complexity. Ordinary blocks built into the environment are used as logic tools. The final design of the logic circuit allows you to study its work in detail at a further stage.

The synthesis of MatLab and Simulink [8, 9] programs serves as a way to design an extensive family of instrumental applications for solving professional problems. The so-called Toolboxes also serve to generate, learn and optimize systems. This can be in demand in various industries when solving various application problems.

2 Materials and Methods

The MatLab system contains a group of functions that allow solving the Koshy problem in systems of ordinary differential equations, given explicitly as dx/dt = F (x, t) and implicitly as M(t, x)dx/dt = F (t, x) - the so-called ODE solvers (solver ODE). They allow the user to select a method, set initial conditions, and other capabilities.

In a simple form, it is enough to apply the command: [T, X] = solver ('fun', [t0, tk], X0), where the values t0 and tk determine the integration spectrum, X0 is the vector of the original values, fun is the name of the function for calculating the right parts of the system, solver is the name of the function used (ode45 - Runge-Kutta method of the fourth and fifth orders, ode23 - the same method of the second and third orders, ode113 - Adams method for the so-called non-rigid systems, ode23s, ode15s - for rigid systems, etc.). Here, rigidity refers to an increased requirement for accuracy - the application of a small step over the entire integration range.

Solver versions are characterized by a certain methodology (by default, relative error 10^{-3} and absolute 10^{-6}), duration, and verified solution. When setting the range of values in the integration spectrum, the number of elements in the arrays corresponds to the accuracy of the solution and with the necessary step.

Let's present the problem for solving in MatLab using solvers of systems of ordinary differential equations (ODE):

Find a solution to the differential equation $y' = x - \cos(y/\pi)$ at the interval [1,7; 2,7], for which the following conditions are specified: y (1, 7) = 5,3.

The solution is made in the MatLab environment. With all this, the ode23 solver is used, using one-step Runge-Kutt methods in the explicit form of the second and 4th orders in the modified sentence of Bogacki and Champin. With the moderate rigidity of the ODE system and reduced accuracy requirements, this method can provide an advantage in solution speed.

The user function g = @ (x, y) [x – cos (y/pi)] is initiated in the Command Window. In the syntax of the function @ (x, y) x is an independent variable, y is a dependent variable, x – cos (y/pi) is the right side of the differential equation.

The decision process is performed by accessing the solver (solver) in Command Window with the following operator: [x, ya] = ode23 (g, [1.7, 2.7], [5.3]).

Plotting in grid form is performed by subsequent operators: plot (x, ya), grid on.

The final view implemented in the MatLab environment is shown in Fig. 1.

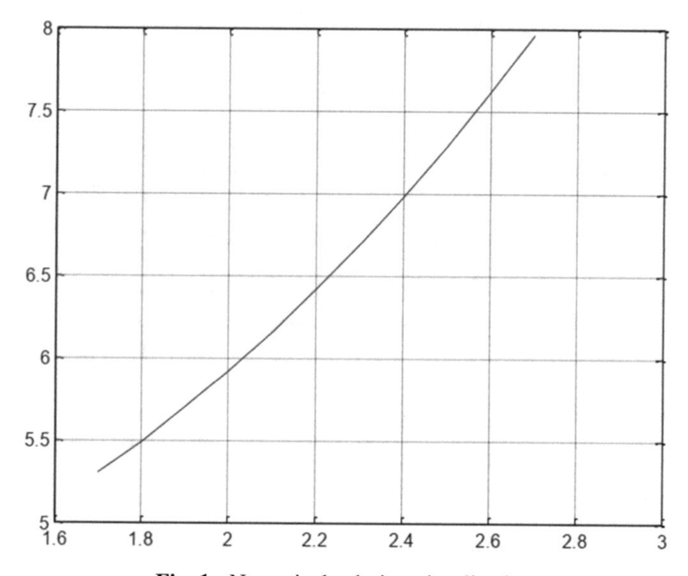

Fig. 1. Numerical solution visualization.

We will formulate another possible problem for solving in MatLab using solvers of systems of ordinary differential equations (ODE).

Find system solution:

$$\begin{cases} x' = -3 \cdot y + \cos t - e^t \\ y' = 4 \cdot y - \cos t + 2 \cdot e^t \end{cases} \tag{1}$$

at initial conditions t0 = 0, x0 = −3/17, y0 = 4/17 using the solver ode 23.

The solution is made in the MatLab environment. The m-file of the function for determining the right halves of differential equations with the name in the editor of the sisdu.m file is created in the editor, then the function can have the following form:

```
function z = sisdu(t, y)
z1 = −3 * y(2) + cos(t) - exp(t);
z2 = 4 * y(2) − cos(t) + 2 * exp(t);
z = [z1;z2];.
```

The following statements are entered in the Command Window:

```
>> t0 = 0; tf = 5; y0 = [−3/17, 4/17];
>> [t,y] = ode23 ('sisdu', [t0, tf], y0);
>> plot(t,y)
>> grid on
```

The final view implemented in the MatLab environment is shown in Fig. 2.

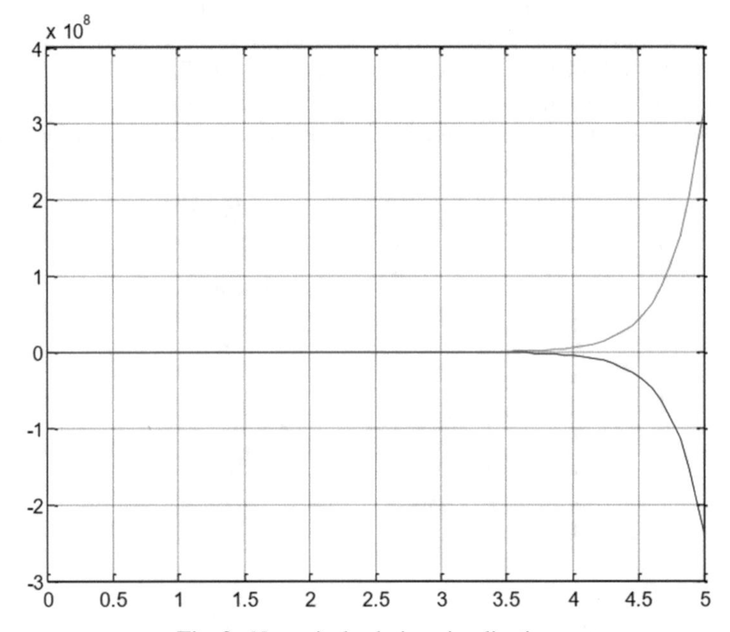

Fig. 2. Numerical solution visualization.

Another such example of an ODE solution requires the creation of the following function:

```
function z = ssisdu(t, y)
a = 0.8; m = 2.7;
z1 = −a * y(1) + a * y(2);
z2 = a * y(1) − (a − m) * y(2) + 2 * m * y(3);
z3 = a * y(2) − (a − m) * y(3) + 3 * m * y(4);
z4 = a * y(3) − 3 * m * y(4);
```

z = [z1; z2; z3; z4];

The following statements are entered in the Command Window:

```
>> [t,y] = ode23 ('ssisdu', [0 1], [1 0 0 0]);
>> plot(t, 100 * y)
>> grid on
```

The final view implemented in the MatLab environment is shown in Fig. 3.

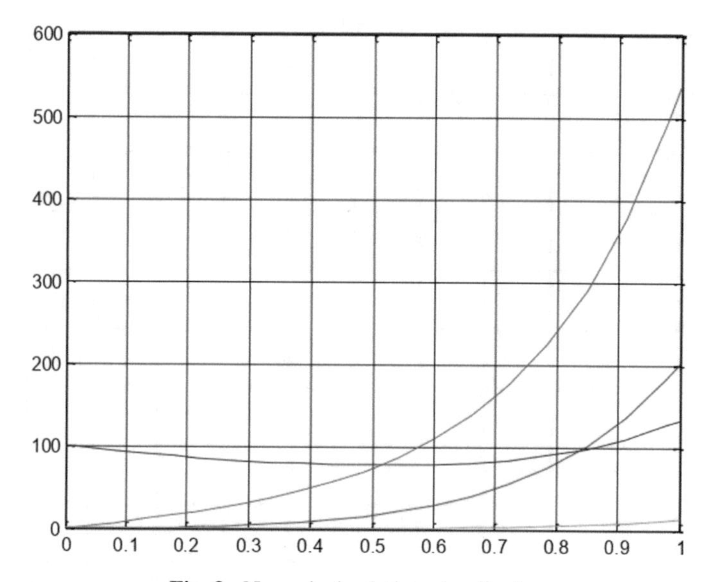

Fig. 3. Numerical solution visualization.

The possibility of solving ordinary higher order differential equations using MatLab solvers is shown on the example of second order ODEs.

Let's solve the second order ODEs $y'' - 2 \cdot y' - y = 6 \cdot x \cdot e^x$ under the given initial conditions: $y(0) = y'(0) = 1$.

At the first step, let's give a differential equation to a system of equations:

$$\begin{cases} y' = y_1 \\ y_1' = 6 \cdot x \cdot e^x + 2 \cdot y_1 + y \end{cases} \tag{2}$$

The solution is made in the MatLab environment. As in the previous examples, you need to create an m-file of the function for calculating the right parts of differential equations.

Let the filename sisdu_3.m, then the function can be as follows:

```
function z = sisdu_3(x, y)
z1 = y(2);
z2 = 6 * x * exp(x) + 2 * y(2) + y(1);
z = [z1; z2];
```

The following statements are entered in the Command Window:

```
>> x0 = 0; xf = 10; y0 = [0, 1];
>> [x, y] = ode23 ('sisdu_3', [x0, xf], y0);
>> plot(x, y(:, 1))
>> grid on
```

The final view implemented in the MatLab environment is shown in Fig. 4.

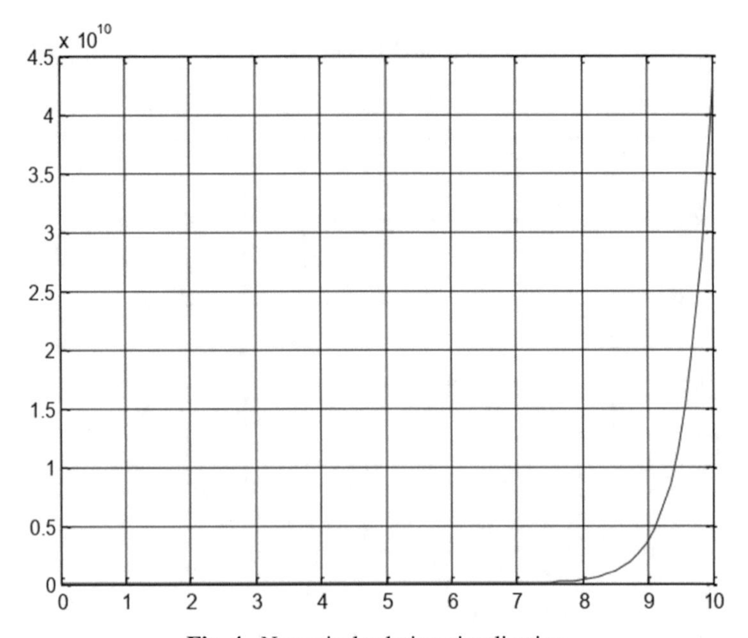

Fig. 4. Numerical solution visualization.

An example of finding a numerical solution to practical problems using MatLab functions is the well-known Lotka-Volterra model of the predator-prey system with a logistic correction described by a system of equations:

$$\begin{cases} x_1^{'} = (a - b \cdot x_2) \cdot x_1 - \alpha \cdot x_1^2 \\ x_2^{'} = (-c + d \cdot x_1) \cdot x_2 - \alpha \cdot x_2^2 \end{cases} \tag{3}$$

with a given number of "victims" and "predators" at the initial moment t = 0. This system of equations is one of the so-called autonomous (or dynamic), where the variable t is obviously not included in the right side of the system. Accordingly, you can not only find a solution x1 = x1(t), x2 = x2(t), but also map their relationship. In a parametric task, the line x1 = x1(t), x2 = x2(t) describes the phase curve (motion line) of the system – a smooth curve without self-intersections, a closed curve, or a point, allowing you to judge the stability of the system to external influences.

The following program, when specifying different α values, creates the corresponding phase portraits (Fig. 5 and 6) - the usual oscillatory process and the gradual death of populations:

```
function f = VolterraLog(x, y)
a = 4; b = 2; c = 2; d = 1; alpha = 0.1;
f(1) = (a − b ∗ x(2)) ∗ x(1) − alpha ∗ x(1)^2;
f(2) = (−c + d ∗ x(1)) ∗ x(2) − alpha ∗ x(2)^2; f = f′;
>> opt = odeset ('OutputSel', [1 2], 'OutputFcn' , 'odephas2');
>> [T, X] = ode45 ('VolterraLog', [0 10], [3 1], opt);
```

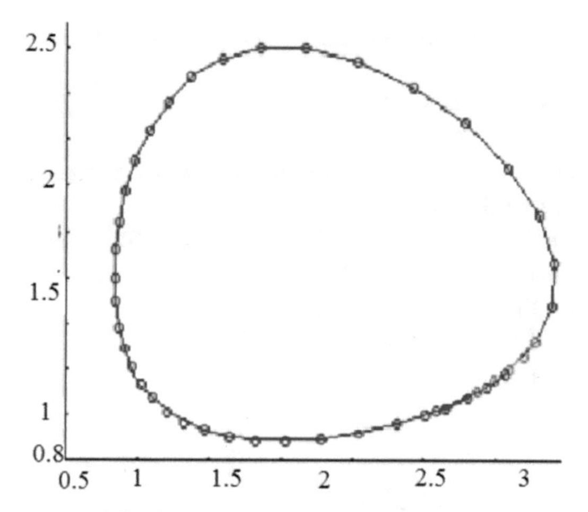

Fig. 5. Phase portrait of oscillations.

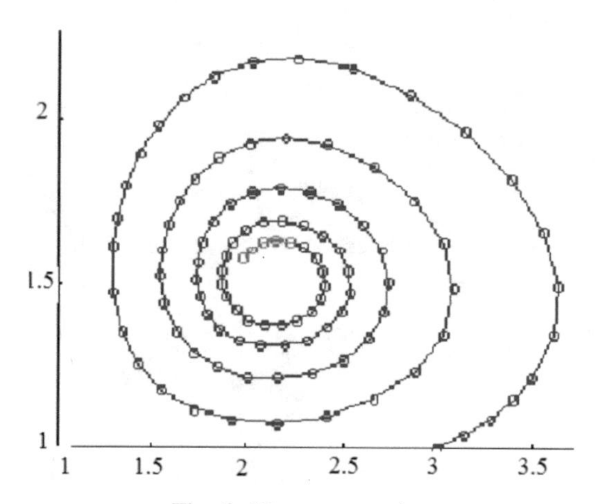

Fig. 6. Phase attenuation.

You can construct a three-dimensional vector field of the system path directions at each point in the phase plane using the odephas3 function. In the particular case, one can consider Euler's three-dimensional problem of moving a solid in space. It has the form:

$$\begin{cases} x_1^{'} = x_2 \cdot x_3 \\ x_2^{'} = -x_1 \cdot x_3 \\ x_2^{'} = -0{,}51 \cdot x_1 \cdot x_2 \end{cases} \quad (4)$$

at x1(0) = x2(0) = x3(0) = 0

The MatLab program of the functions on the right side has the form:

function f = Eiler(t, x).

f(1) = x(2) * x (3); f(2) = −x(1) * x (3);

f(3) = −0.51 * x (1) * x(2); f = f′;

Problem solving operators: ode45 for calculating and constructing a portrait, plot for visualizing solutions at a specific step.

```
>> opt = odeset ('OutputSel', [1 2 3], 'OutputFcn', 'odephas3');
>> [T,X] = ode45 ('Eiler', [0 7.25], [0 0 1], opt);
>> plot(T, X)% Fig. 7
>> [T,X] = ode45 ('Eiler', [0:0.25: 7.25], [0 1 1]);
>> plot(T, X)% Fig. 8
```

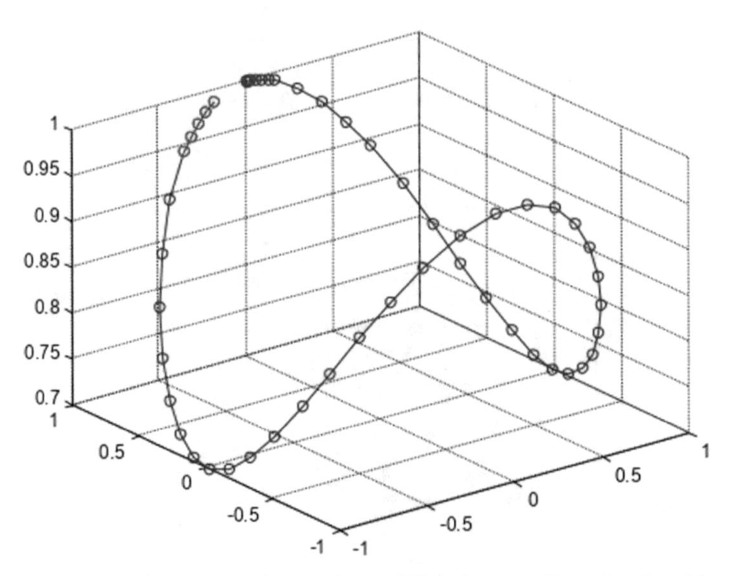

Fig. 7. Three-dimensional phase portrait of Euler's three-dimensional problem.

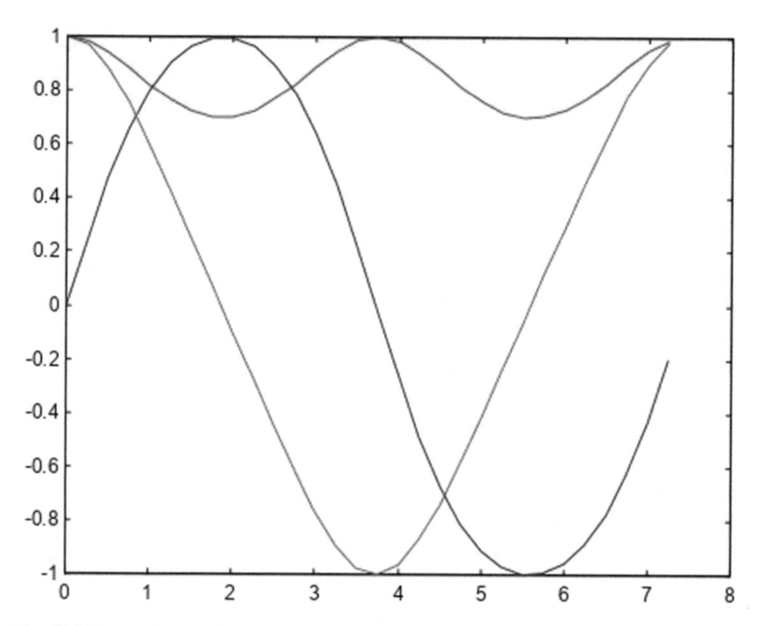

Fig. 8. Three-dimensional phase portrait of Euler's three-dimensional problem.

Solving the problem of numerical integration essentially boils down to computing a particular integral. A numerical solution can be obtained in a variety of ways. In a number of technically practical problems, it is necessary to find a solution to a certain integral of type:

$$J[f] = \int_a^b f(x)dx \tag{5}$$

Only in selected versions it is possible to find the antiderivative F(x) function f(x) and represent the integral in the form [10, 11]:

$$\int_a^b f(x)dx = F(b) - F(a) \tag{6}$$

Also, if the integrator function is not analytically specified, it becomes necessary to use numerical integration methods.

The problem of numerical integration [12, 13] consists in determining the approximate value of the integral. A solution to such a problem is possible using the formula for approximate calculation of the integral by the value of the integrally function:

$$J_N[f] = \sum_{i=1}^{N} c_i \cdot f(x_i) \tag{7}$$

where $f(x_i)$ are the values of the function f (x) at node points $x = x_i$, c_i are the values of weight factors (weights) that depend only on nodes, but do not depend on the choice of specific (x).

Formula (7) corresponds to the quadrature formula [14–16].

1) Rectangle method:

   - Method of left rectangles
   - Right rectangle method
   - Middle rectangle method

   Of the methods of numerical integration according to the methods of rectangles, the smallest in terms of error is the simplest quadrature formula of central or middle rectangles. In practice, this can be proved by using Taylor series expansion.

   Calculated dependency for middle or other central rectangles:

$$\int_a^b f(x)dx \approx h\sum_{i=1}^{n}\left(x_{i-1} + \frac{h}{2}\right) \tag{8}$$

2) Method of numerical integration by trapezoid method:

$$\int_a^b f(x)dx \approx h\sum_{i=1}^{n}\left(\frac{f(x_0 + f(x_n)}{2} + \sum_{i=1}^{n-1} f(x_i)\right) \tag{9}$$

3) Parabol numerical integration method (Simpson) [17]. For three integration points, we have:

$$\int_a^b f(x)dx \approx \frac{b-a}{6}\left(f(a) + 4f\left(\frac{a+b}{2}\right) + f(b)\right) \tag{10}$$

3 Results AHD Discussion

Let's define the initial data of the problem for solving in MatLab using simple quadrature formulas.

It is necessary to determine the integral of the function f (x) = ln (x + 1) *dx* with a partition step h = 0.01 over a range of values from 0 to 1.

An analytical solution has been defined for this task:

$$\int_0^1 \ln(x+1)dx = ((x+1)\cdot\ln(x+1)-x)|_0^1 = 2\ln 2 \approx 0{,}38629436... \quad (11)$$

The program was executed in the MatLab software environment, and the program listing is shown in Fig. 9.

In the final implementation of the written code in the program, it was possible to get (Table 1) approximate integral values using a family of different numerical integration methods, as well as absolute deviations of the resulting solution compared to the exact one.

Table 1. Program results.

Step	Implementation method					
	Central rectangles		Trapezes		Simpson	
	Value interval	Deviation	Value interval	Deviation	Value interval	Deviation
0.01	0.3862902	$4.159 \cdot 10^{-7}$	0.37935	$6.5 \cdot 10^{-4}$	0.3862943	$9.64 \cdot 10^{-12}$
0.1	0.385878	$4.159 \cdot 10^{-5}$	0.3197136	$6.36 \cdot 10^{-3}$	0.3862934	$9.64 \cdot 10^{-12}$

In addition, during the software implementation of the tasks discussed, MatLab measured the speed of calculating the integral by various methods. The following results were established: 0,000611 c (middle rectangles); 0,000655 c (trapezoids); 0.000681 c (Simpson). The established indicators demonstrated that in reality, the error order of the middle rectangle method is $1 \cdot (O\ (h))$, the trapezoid method is $2 \cdot (O\ (h^2))$, and the Simpson method corresponds to $4 \cdot (O\ (h^4))$.

```
clc
clear
close all
I=vpa(2*log(2)-1);%the exact value of the integral
F=@(x) (log(x+1)); % specifying the integrative function
a=0;b=1; %integration interval
h=0.01; %setting the step
x=a:h:b;
sum=0;
for i=2:length(x)-1 %calculation using the method of middle rectangles
sum=sum+F(x(i-1)+h/2);
end
pr=vpa(h*sum);
s=h*(F(x(1))+F(x(end)))/2;
for i=2:length(x)-1
s=s+h*F(x(i));
end
trap=vpa(s); %trapezoid method
s1=F(x(1))+F(x(end));
sum1=0;sum2=0;
for i=1:(length(x)-1)/2;
sum1=sum1+F(x(2*i-1));
sum2=sum2+F(x(2*i));
end
J=vpa((s1+4*sum2+2*sum1)*h/3); % Simpson's method
disp('Mean rectangle integral value: ');disp(pr);
disp('Absolute error: ');disp(abs(pr-I));
disp('Trapezoidal integral value: ');disp(trap);
disp('Absolute error: '); disp(abs(trap-I));
disp('Simpson integral value (parabol): ');disp(J);
disp('Absolute error: '); disp(abs(J-I));
```

Fig. 9. MatLab program listing.

4 Conclusions

Summing up the work, the article discusses the problems of solving ordinary differential equations (ODE) in MatLab using the corresponding equations of the first and highest orders of ODE, as well as ODE systems.

Based on the results of numerical integration (Table 1), it is obvious that from the data of 3 simple methods, based on accuracy considerations, at the same step, the Simpson method (parabol) is. The calculation time for each method is approximately the same. These methods, without taking into account the relatively small order of accuracy, can be used for a large part of practical calculations and can be one of the practical tools for use in the so-called Industry 4.0 [18].

In general, MatLab is a unique collection of implementations of modern numerical methods of computer mathematical operations that have been created over the past three decades. She included as experience, techniques and extensive knowledge formed during thousands of years of evolution of the science of mathematics as such. The system and software complex has a wide documentation base in electronic form both for large-scale individual computers and for super computers.

The boundaries of using MatLab are wide, including due to the high speed of implementation of tasks by the built-in solver. Due to the presence of an extended catalog of mathematical operations, MatLab is widely used in problems of mathematical modeling of mechanical systems used as an object of study in problems of dynamics, hydrodynamics, aerodynamics, acoustics, energy and other problems.

The MatLab function library, combined with excellent 2D or even 3D graphics, allows you to solve the problems of various areas of research. An example of similar tasks is the problem of population dynamics considered in the article, created for biological systems, but with certain adjustments is applicable to company competition, the construction of financial pyramids, population growth, the spread of epidemics, overcrowding, nuclear energy and extensive environmental problems.

References

1. Valentine, D., Hahn, B.: Essential MATLAB for Engineers and Scientists, 3rd edn. Elsevier, Amsterdam (2022)
2. Dyakonov, V.P.: MatLab. Full Self-teacher Earing Theory. DMK-Press, Moscow (2012)
3. Kurasov, D.: Calculation and design of mechanical transmissions with the help of various computer-aided design in conditions of educational process. In: IOP Conference Series: Materials Science and Engineering, p. 032098 (2020)
4. Bebikhov, Y., Semenov, A., Yakushev, I., Kugusheva, N., Pavlova, S., Glazun, M.: The application of mathematical simulation for solution of linear algebraic and ordinary differential equations in electrical engineering. In: IOP Conference Series: Materials Science and Engineering, vol. 643, no. 1, p. 012067 (2019)
5. Arjunan, R., Sarala, S., Nirmala, M.: Matlab solution for first order differential equations on real time engineering applications. J. Adv. Res. Dyn. Control Syst. **10**(7), 202–209 (2018)
6. Hansen, P.: Regularization tools: A Matlab package for analysis and solution of discrete ill-posed problems. Numer. Algorithms **6**(1), 1–35 (1994)
7. Lee, K., Comolli, N., Kelly, W., Huang, Z.: MATLAB-based teaching modules in biochemical engineering. Chem. Eng. Educ. **49**(2), 95–100 (2015)
8. Lee, K., Comolli, N., Punzi, V., Kelly, W., Huang, Z.: Teaching tip: results of a survey on MATLAB & MathCAD education in bio-chemical engineering. Chem. Eng. Educ. **48**(1), 59 (2014)
9. Semenova, M., Vasileva, A., Lukina, G., Popova, U.: Solving differential equations by means of mathematical simulation in simulink app of MATLAB software package. In: Mottaeva, A. (ed.) Technological Advancements in Construction. LNCE, vol. 180, pp. 417–431. Springer, Cham (2022). https://doi.org/10.1007/978-3-030-83917-8_38
10. Makarova, E., Gudkova, E.: Modern information technology. **27**, 17–22 (2018)
11. Zhang, Z., Karniadakis, G.: Numerical Methods for Stochastic Partial Differential Equations with White Nois. Springer, Heidelberg (2017). https://doi.org/10.1007/978-3-319-57511-7
12. Li, L.: Partial differential equation calculation and visualization. J. Discrete Math. Sci. Cryptogr. **20**(1), 217–229 (2017)
13. Gimeno, J., Jorba, A., Jorba-Cuscó, M., Miguel, M., Zou, M.: Numerical integration of high-order variational equations of ODEs. Appl. Math. Comput. **442**, 127743 (2023)

14. Ali, A., Abbas, A.: Applications of numerical integrations on the trapezoidal and simpson's methods to analytical and MATLAB solutions. Math. Modell. Eng. Probl. **9**(5), 1352–1358 (2022)
15. Oberbroeckling, A.: Programming Mathematics Using MATLAB®. Academic Press, Cambridge (2021)
16. Hayotov, A.R., Bozarov, B.I.: Optimal quadrature formula with cosine weight function. Probl. Comput. Appl. Math. **4**, 106–118 (2021)
17. Zhao, Z., Jiang, B.: The application of subsection simpson integration on artillery firing efficiency simulation modeling. Appl. Mech. Mater. **2655**, 130–134 (2011)
18. Kurasov, D.: Computer-aided manufacturing: Industry 4.0. In: IOP Conference Series: Materials Science and Engineering, vol. 1047 (2021)

Creation of an Intelligent Automated Business Process Management System Through Speech and Dialogue Design Patterns

Victoria Shamraeva[1(✉)], Valery Abramov[1], Irina Nikolaeva[2], Yanming Wang[3], Yaroslav Zubov[1,4], Marina Medvedeva[1], and Dmitry Morkovkin[1]

[1] Financial University under the Government of the Russian Federation, 49/2, Leningradsky Avenue, Moscow 125167, Russia
VVShamraeva@fa.ru
[2] Samara State University of Economics, 141, Sovetskoi Armii St, Samara 443090, Russia
[3] Shenzhen University, 3688, Nanhai Ave, Shenzhen, Guangdong, China
[4] Russian State University for the Humanities, 6, Miusskaya Sq, Moscow 125047, Russia

Abstract. The subject of the study is the technical and organizational system of working with knowledge in a large organization. The object of the study is a large organization with a large flow of information. The goal is to develop algorithms for finding the fastest way to find dialogue patterns, given a large amount of input data. On the basis of such algorithms, to develop recommendations for the creation of an intelligent automated control system in a large organization. When building search queries to create a knowledge base, the ideology of ontological modeling is used. The proposed algorithms will allow a large organization with unique business processes to extract the necessary information from the knowledge base and make decisions in the shortest possible time. The created ontological model and recommender system, built on the basis of speech and dialogue patterns, will not only reduce the waiting time for receiving a request, but will also give the most optimal solution. This will reduce the risk of making the wrong decision and the dissatisfaction of potential external participants and members of the organization, which is certainly an economically beneficial factor for it.

Keywords: Patterns · Ontology · Ontological Design Patterns · Design Patterns · Clustering

1 Introduction

A pattern is a repeating construction of images, phrases, actions, etc. Pattern detection is highly relevant. Obviously, the same phenomenon under the same conditions will lead to the same result and, as a result, the timely detection of such a dependence will minimize damage or maximize profit from the result obtained. Also, the very fact of the repetition of a particular phenomenon can be used for further work on the data. Having a systematic catalog of patterns, you can use it when designing similar classes of problems [1, 2]. Initially, such a catalog of patterns was developed by Alexander Christopher in

© The Author(s), under exclusive license to Springer Nature Switzerland AG 2023
A. Gibadullin (Ed.): DITEM 2022, LNNS 683, pp. 15–22, 2023.
https://doi.org/10.1007/978-3-031-30926-7_2

the late 1970s [3, 4] and was used only for the design of buildings and cities. Finding patterns seems like a trivial task with a small amount of data and simple patterns. For example, in a geometric wallpaper pattern, finding patterns is not difficult. However, assuming there is a kilometer by kilometer picture that contains more than a hundred possible patterns repeating chaotically, searching for these patterns without the use of computer processing will obviously take a large number of man-hours. For software development, patterns began to be used decades later [5–7], then the concept of pattern programming in various programming languages [8–11] appeared, and the need for its application in various areas of life: optimization of business processes, development of various learning systems, etc. [12–15]. Programs and algorithms for searching for patterns are more common in everyday life than it seems: for example, a bank can block a client's card if a suspicious transaction was made on the card. In this case, a "suspicious" transaction can be considered a pattern, the search for which in a variety of client transactions allows you to secure his finances.

In this article, the main focus is on the search for speech and dialogue patterns. Speech patterns are repeated words or phrases, dialogue patterns are repeated parts of some dialogue. For example, on a greeting, you will most often receive a return greeting, this is the dialogue pattern. At this stage, it is necessary to identify a clear interaction of intelligent systems between themselves and a person, which is a section of artificial intelligence - the ontology of design. The detection of speech patterns is carried out using substring search algorithms in a string, which, depending on the complexity of the algorithm, go through the entire text in various ways and find the desired lines in it. The problem with such algorithms when searching for speech patterns is as follows: you should know in advance which speech structures should be searched for. With a large amount of data, this problem transforms into the problem of time spent: it is not difficult to divide a relatively small passage of text into words and check how often they occur in a given text, but the speed of such algorithms is quite low. That is, with a large amount of data and the absence of a starting point, upper levels of ontologies, patterns to be searched for, the operation of the algorithm, at best, will take a huge amount of time, and at worst, a huge amount of memory for intermediate iterations. In the case of searching for patterns of dialogue, the task becomes even more difficult, since the same pattern can sound and look different: the words "hello", "aloha", "good afternoon" and so on can be considered as a greeting.. If at the stage of searching for a speech pattern a large number of different variations appear, then at the level of a dialogue pattern there are many more of them, and all these variations must be processed by solving problems of ontological modeling. The development of an ontological model and algorithms for creating a recommender system based on speech and dialogue patterns should reduce the waiting time for receiving a response to a request and extracting the most optimal answer from the knowledge base. This reduces the risk of making the wrong decision and the dissatisfaction of potential external participants and participants in the business structure of the organization, which is certainly an economically beneficial factor for it.

2 Subject, Tasks and Methods

The object of the study is a large organization with a large flow of information and the creation of an intelligent automated control system in it based on speech and dialogue patterns. The purpose of this study is to find the fastest way to find dialogue patterns, given a large number of input data. At the same time, the main problem is the minimum of any starting points - as starting points we have a limited number of keywords, the search for which will help only in data markup, but not in the search for dialog patterns. The main algorithms that were used in the study:

- Substring search algorithms in a string;
- Naive algorithm;
- The Knuth-Morris-Pratt algorithm.

Let's dwell briefly on each of them.

In algorithms for searching for a substring in a string, as data, we have a string or text in which the substring should be found. Such algorithms differ in the speed of work, degree and methods of preliminary marking of data.

The naive algorithm is called naive because it does not imply any extra actions, except for checking the presence of a substring in a string character by character. If the substring does not match the one being compared, the substring is shifted by one character (Table 1).

Table 1. Naive Algorithm.

Step 1													
a	v	t	o	k	o	t	o	c	t	r	o	f	a
k	o	t											
Step 2													
a	v	t	o	k	o	t	o	c	t	r	o	f	a
	k	o	t										
Step 3													
a	v	t	o	k	o	t	o	c	t	r	o	f	a
		k	o	t									
Step n													
a	v	t	o	k	o	t	o	c	t	r	o	f	a
				k	o	t							

The problem with this method is its speed: until the algorithm finds a complete match, every substring in the string will be checked. If, as in the above example, when finding a substring, the algorithm stops, then in the case of the task, in some cases it is necessary to search for a substring in the entire text, that is, check each substring character by character. Also in this example, if the first character does not match, the algorithm immediately shifts. If we assume that the string has the form "aaaaaaaaaaaaaaaaab", and the desired substring has the form "aaaaaaa", then after each shift, the algorithm will have to compare the entire string instead of one character to go to the next shift. However, this algorithm has an unobvious advantage: it does not require any more iterations or allocate memory to store any data. In some situations, this speeds up the search process compared to other algorithms.

The Knuth-Morris-Pratt algorithm differs from the naive algorithm in that it allows you to shift a substring by several characters at once, depending on the prefix function that is formed at the preparation stage. A prefix is the beginning of a substring, and a suffix is its end. At the same time, the same string can have several prefixes and suffixes depending on its length, for example, the string "cat" has the prefixes "c" and "ca" and the suffixes "t" and "at". Let's consider an example of forming a prefix function for each character of the "abcabcabcd" substring. For the character n, the prefix function will return a value equal to the maximum length of the matching prefix and suffix in the substring that ends with that character (Table 2).

Table 2. An example of forming a prefix function for each character of the substring "abcabcad".

n	0	1	2	3	4	5	6	7
Symbol	a	b	c	a	b	c	a	d
Prefix function								

For a character n = 0, the prefix function is 0 because the prefix or suffix cannot equal the entire string. Similarly for characters 1 and 2: the prefix of character 1 - "a", the suffix - "b" do not match, similarly for character 2, where the suffixes are "a" and "ab", and the prefixes "c" and "bc" (Table 3).

Table 3. An example of forming a prefix function for each character of the substring "abcabcad".

n	0	1	2	3	4	5	6	7
Symbol	a	b	c	a	b	c	a	d
Prefix function	0	0	0					

Find the function prefix for each character (Table 4).

Thus, as a result of preparation, we get the following table (Table 5).

Consider the iterations of the algorithm, for convenience, the numbers of symbols are signed under the required substring (Table 6). At the first step, we see that character 4 did not match the original string. In this case, we focus on the value of the prefix function of character 3. It is equal to 1, which means we shift the substring so that character 1 is in place of character 4. Now character 1 does not match the original string, the prefix function of the previous character is 0, which means we shift the substring so that the character 1 is replaced by the character 0. With such simple iterations, we reach the desired substring, skipping some of the iterations that a naive algorithm would have performed. In the example where the naive algorithm goes through many comparisons, the Knuth-Morris-Pratt algorithm will step over some of the values according to the prefix function.

Table 4. Prefix feature for each character.

Symbol	Prefixes	Suffixes	Prefix function
a	-	-	0
b	a	b	0
c	a, ab	c, bc	0
a	a, ab, abc	a, ca, bca	1(a)
b	a, ab, abc, abca	b, ab, cab, bcab	2(ab)
c	a, ab, abc, abca, abcab	c, bc, abc, cabc, bcabc	3(abc)
a	a, ab, abc, abca, abcab, abcabc	a, ca, bca, abca, cabca, bcabca	4(abca)
d	a, ab, abc, abca, abcab, abcabc, abcabca	d, ad, cad, bcad, abcad, cabcad, bcabcad	0

Table 5. An example of forming a prefix function for each character of the substring "abcabcad".

n	0	1	2	3	4	5	6	7
Symbol	a	b	c	a	b	c	a	d
Prefix function	0	0	0	1	2	3	4	0

In addition to the mentioned naive algorithm and the Knuth-Morris-Pratt algorithm, there are many algorithms that optimize the search for a substring in a string. The question of using one of them as an auxiliary method for finding dialogue patterns can be considered a question of what to look for and why. In the case of the data in question, a search is made on the keywords of the dialogues to find where and in what context they occur. However, the definition of dialogue patterns is impossible by any of the proposed methods - they require an exact match of a substring with a string, which is impossible. In the presence of softer methods of searching and comparing, another problem arises: there are no initially found patterns that should be tracked, that is, there are no input data for the algorithm. Thus, the use of softer comparison methods becomes impossible, and the use of substring search algorithms in a string becomes meaningless.

One of the development methods in this study is the possibility of data clustering. Clustering is a way of combining many different objects into groups based on the characteristics of these objects. Previously, this concept was associated with the processing of anthropological data. For example, a group of people can be grouped into clusters by age, hobbies, earnings, and so on. At the same time, different clustering methods are based on different criteria and algorithms, due to which the same data set can be divided into a different number of clusters and in completely different ways, so the correct determination of the number of clusters and the algorithm is critical. Clustering will

Table 6. Algorithm iterations.

Step 1	a	b	c	a	d	c	a	b	c	a	b	c	a	b	c	a	d
	a	b	c	a	b	c	a	d									
	0	1	2	3	4	5	6	7									
Step 2	a	b	c	a	d	c	a	b	c	a	b	c	a	b	c	a	d
				a	b	c	a	b	c	a	d						
				0	1	2	3	4	5	6	7						
Step 5	a	b	c	a	d	c	a	b	c	a	b	c	a	b	c	a	d
							a	b	c	a	b	c	a	d			
							0	1	2	3	4	5	6	7			
Step 6	a	b	c	a	d	c	a	b	c	a	b	c	a	b	c	a	d
										a	b	c	a	b	c	a	d
										0	1	2	3	4	5	6	7

allow you to identify groups of clients that are similar to each other - it is obvious that in different clusters of clients dialogue patterns inherent in this type of client will be found. Such a division will allow creating a compact selection of dialogs that matches the entire amount of data, so that it becomes possible to process dialogs with minimal use of substring search algorithms in a string, almost manually.

However, regardless of the input data, there are a number of anomalies in the data - outliers. Consider methods for finding outliers. In the context of the available data, for example, a call from a client with a zero turnover on the card and a request for a deposit of more than a billion rubles will be considered an outlier. If such outliers are not filtered, the clusters will contain data that actually does not belong to these clusters, since most algorithms are aimed at clustering all objects without exception. However, there are algorithms that are not quite suitable in this case as clustering, but they highlight outliers. One example is the DBSCAN algorithm, clustering based on object density and its modification. It has two parameters - the radius around the point and the number of points that must be included in this radius so that the point can be attributed to the cluster. If there are n points around point A within radius r, then point A belongs to this cluster. Similarly, all points that were within the radius of point A are checked, and so on, until the cluster is fully formed.

This algorithm is not quite suitable as a clustering algorithm, since it does not determine nested clusters, considering them to be one large cluster. Also, if several clusters have different densities, the algorithm can consider points that actually belong to one of the clusters as outliers. Nevertheless, for filtering obvious outliers, the algorithm fits perfectly, including because of the simple implementation.

The group of clustering algorithms that can suit the task at hand is called hierarchical clustering. Neural networks related to non-hierarchical algorithms are not fully implemented here. There are two types of algorithms: agglomerative and divisive. Agglomerative clustering algorithms single out each sample object as a cluster and then combine clusters according to various criteria, while divisive ones, on the contrary, consider all points to be objects of one cluster, and then divide it into parts. This group of algorithms has its own method for determining the number of clusters: building a dendrogram - a graph that visually represents the order in which clusters were combined or divided. Agglomerative methods are divided according to the criterion of combining clusters: distance of nearest neighbors, distance of far neighbors, group average distance. Each of the methods has its advantages and disadvantages; the choice of algorithm should be based on the input data.

Regardless of the agglomerative method, there is a problem of stopping the algorithm when a certain number of clusters is reached. Obviously, too few of them will lead to generalization of data that should not be generalized, and too much of them will lead to excessive specification and the appearance of insignificant clusters.

In the case of hierarchical methods, there is a great way to determine the optimal number of clusters, which also allows you to track the order in which clusters are combined - building a dendrogram. The number of clusters can be determined both automatically by stopping the algorithm at a certain one, and based on the constructed dendrogram, decide how many clusters should be specified.

There is also a universal method for determining the number of clusters, regardless of the algorithm - the elbow method. When applying this method, it is required to build an algorithm for several options for the number of clusters and the sum of intracluster distances for each of the options. Obviously, the more clusters, the closer the intracluster distance will tend to zero. The task is to independently determine from the graph the dependence of the intracluster distance on the number of clusters, when an increase in the number of clusters will not give a significant reduction in the intracluster distance.

3 Method Development

As a correct solution, it is proposed to choose, not by chance, a part of the dialogues, which are then manually, using the algorithms listed in the previous paragraph, parsed into patterns. It is assumed that as a result we will get a selection of dialogs, which includes all possible types of requests and patterns. If you select dialogs randomly, there is a risk of skipping many patterns, so before choosing, we will divide all calls into groups, and randomly select dialogs from different groups, according to their sizes.

There are two problems in clustering any kind of data: the parameters by which clustering should be carried out and the method of clustering. As parameters by which it is possible to carry out clustering, one can single out: age, gender, turnover on the client card, products purchased by the client, credit history, whether the client is a client of the bank, the main request of the client, the time of the call, etc. The fewer clustering parameters will be used, the clearer clusters can be identified.

4 Conclusion

The proposed algorithms will allow a large organization with unique business processes to extract the necessary information from the knowledge base and make decisions in the shortest possible time. The construction of an ontological model based on such algorithms will be part of the intelligent automated system of the entire organization as a whole. The relevance of the ontological model itself depends on the specific business processes currently taking place in the organization and requires constant updating.

References

1. Vlissides, J.: Pattern Hatching. Design Patterns Applied Reading [etc.]. Addison-Wesley (1998)
2. Gamma, E., Richard, H., Ralph, E., Vlissides, J.: Design Patterns: Elements of Reusable Object-oriented Software. Addison-Wesley Professional Computing Series. Reading (1994)
3. Alexander, S., Ishikawa, S., Silverstein, M.: A collection of patterns which generate multi-service centres. In: Declan, Kennedy, M., (eds.). The Inner City. Architects Year Book 14, Elek, London (1974)
4. Alexander, S., Ishikawa, S., Silverstein, M.: A Pattern Language: Towns, Buildings, Construction. Oxford University Press, Oxford (1977)
5. Buschmann, F., Meunier, R., Rohnert, H., Sommerlad, P., Stal, M.: Pattern-oriented software architecture. Syst. Patterns **1**, 476–480 (1996)
6. Teplyakov, S.V.: Dependency injection patterns. RSDN Mag. **4**, 14–22 (2012)
7. Krainova, E.A.: Theoretical aspects of pattern programming. Bull. Volga Univ. V.N. Tatishchev **2**(21), 82–90 (2013)
8. Feshina, E.V., Omelchenko, D.A., Gonataev, R.G.: Implementation of design patterns in the PYTHON programming language. Innov. Sci. Educ. **28**, 983–987 (2021)
9. Gibadullin, A., Pulyaeva, V.: Obstacles to the formation of a common electricity market of the Eurasian Economic Union. In: E3S Web of Conferences, vol. 114, p. 02002 (2019)
10. Dorodnykh, N.O., Yurin, A.Yu.: Ontology design patterns engineering based on analysis and transformation of concept maps. In: Proceedings of the Eighth International Conference, pp. 153–161 (2019)
11. Shamraeva, V.V., Chuvikova, V.V.: Web based supply chain business process optimization. Self Manage. **2**(119), 628–632 (2020)
12. Romanova, Ju.A., Morkovkin, D.E., Romanova, Ir.N., Artamonova, K.A., Gibadullin, A.A.: Formation of a digital agricultural development system. In: IOP Conference Series: Earth and Environmental Science, vol. 548, p. 032014 (2020)
13. Orazbayev, B., Ospanov, E., Kissikova, N., Mukataev, N., Orazbayeva, K.: Decision-making in the fuzzy environment on the basis of various compromise schemes. Proc. Comput. Sci. **120**, 945–952 (2017)
14. Chertkova, E.A.: Application of design patterns to the development of computer-aided learning systems. Bull. Tambov State Tech. Univ. **13**(1), 13–19 (2007)
15. Hazieva, A., Rafikova, N., Habirov, G., Zalilova, Z., Sagadeeva, E.: Econometric models of cattle-breeding production cost. Industr. Eng. Manage. Syst. Link Disabled **19**(4), 857–865 (2020)
16. Orazbayev, B.B., Ospanov, Y.A, Orazbayeva, K.N., Serimbetov, B.A.: Multicriteria optimization in control of a chemical-technological system for production of benzene with fuzzy information. Bull. Tomsk Polytech. Univ. Geo Assets Eng. **330**(7), 182–194 (2019)

Justification of the Use of Mathematical Analogues in the Construction of an Optimization Model of the Company's Functioning, Taking into Account Its Economic Security

M. I. Korolev[1], D. D. Lapshin[2], M. L. Lapshina[3], A. I. Khorev[3], and O. O. Lukina[3](✉)

[1] Regional Branch Association of Employers of the Moscow Region in the Field of Safety and Security "Global-Security", Moscow, Russia
[2] Municipal Budgetary Educational Institution "Novousmansky Educational Center", Novaya Usman, Voronezh, Russia
[3] Voronezh State Forestry Engineering University named after G.F. Morozov, Voronezh, Russia
oks.lukina@gmail.com

Abstract. The paper offers one of the options for constructing a simulation model of the company's functioning, taking into account optimization approaches in the implementation of a number of technological processes. The possibility of adapting linear programming tasks to the adoption of the most optimal management decision focused on the economic security of the company, as well as objective prerequisites for changing technological parameters, from the range of acceptable values and consequences leading to a change in the characteristics of the relevant services and products is considered. Also, an approach is proposed to construct a solution to the convex programming problem, interpreted as finding the optimum of production planning. The analysis of potentially possible modes of self-organization of a technical system is proposed, taking into account the use of the utility function.

Keywords: Resources · Volumes · Trajectory · Function · Technology · System

1 Introduction

It should be noted that systems theory played a huge role for the development of management theory and systems of management decision support [1]. Its application to socio-economic and information systems made it possible to create a unified research institute for various fields of economic knowledge.

The system approach is based on interdisciplinary research of any problem from the point of view of general patterns of development, adequately specified in relation to the specifics of the problem under consideration. At the same time, the attention of the researcher should be focused on identifying the deep basis of the development of the

© The Author(s), under exclusive license to Springer Nature Switzerland AG 2023
A. Gibadullin (Ed.): DITEM 2022, LNNS 683, pp. 23–31, 2023.
https://doi.org/10.1007/978-3-031-30926-7_3

system under study and revealing the mechanisms of its activity within the dynamics of existing intra-system and non-system connections [2].

An important feature of the system is the predominance of intra-system connections of its components over external influences on them. That is why the most effective method of managing large systems is management through mechanisms of self-regulation and self-control. In this case, a self-regulating system can be defined as a complex technical system capable of maintaining or improving its organization depending on changes in external and internal conditions. The need for self-adjustment and self-regulation is especially relevant with changing environmental parameters. Since the process of developing and implementing any system has a certain time lag, a system optimized for specific conditions that existed at the time of its development will not necessarily be effective if these conditions change by the time of its implementation.

2 Initial Data for Solving the Optimization Problem

Let's consider the initial formulation of the optimization problem in the following form: we assume x is the vector of services, R is the vector of resources, we write the following type of utility function

$$u(x, \alpha) = \sum_{j=1}^{n} \alpha_j u_j(x_j) \rightarrow \max \tag{1}$$

with corresponding resource constraints $g(x)$

$$g(x) = \sum_{j=1}^{n} g^{(j)}(x_j, \beta^{(j)}) \leq R, \tag{2}$$

$$x \in Q, \tag{3}$$

where $j = 1, \ldots, n$ – defines the corresponding service; $R = (R_1, \ldots, R_m)$ – composing vectors of resources, $i = 1, \ldots, m$ – corresponds to the index of the resource; $u_j(x_j)$ – monotonically increasing utility function of the production of the j – th service x_j, $w(x, \alpha)$ – determines the utility function of the "consumer basket" of j-th services; α_j – weight services j in the "consumer basket" of j-th services; $\beta^{(j)}$ – a vector strictly convex function of resource consumption when providing the j–th service x_j, $\beta^{(j)}$ – a vector of parameters corresponding to the j-th technological process $\beta^{(j)} \in B^{(j)}$, $j = \overline{1, n}$ – defines convex sets. We consider the weights α_j to be constant values equal to 1.

We assume that some j-th service is produced in accordance with the j-th technological process and, conversely, the final result of the work of the j-th technological process is the j-th service. Due to the interconnectedness of the processes of providing services and the corresponding technological processes, we will use the concepts of specifically j-th technology and the corresponding j-th service alternately, the order is determined by the specific characteristics of the technological process or the provision of a certain service.

We will assume that the technologies of the corresponding services are dependent on some parameters $\beta^{(j)} \in B^{(j)}$ belonging to the scope of acceptable parameters $B^{(j)}$. The variation of these parameters leads to a change in the characteristics of j technologies, in particular resource intensity, direct, overhead, fixed and variable costs. If the opposite is not specifically discussed, we assume that the technology parameters are constant: $\beta^{(j)} = const$.

Let be the solution of the problem (1)–(3), and is measured as a set of provided services. Consider the vector p:

$$p_j = \partial u(x^{(*)})/\partial x_j \tag{4}$$

and the vector w:

$$w_j = \partial u(x^{(*)})/\partial R_j. \tag{5}$$

In this case, vector p turns out to be a vector of prices for services, and vector w is a vector of prices for resources. Indeed, at point $x(*)$, the surface of the constant level $u(x) = u(x^{(*)}) = const$ of the utility function $u(x)$ touches the surface of the constant level of the resource consumption function $g(x) = g(x^{(*)}) = const$. Therefore, the plane $(p,\ x) = (p,\ x^{(*)}) = const$ is a set of equivalent exchanges of services. The number of equivalent exchanges that do not change the utility function in the first approximation is described by the equation

$$\sum_{j=1}^{n} p_j \Delta x_j + \sum_{i=1}^{m} w_i \Delta R_i = 0, \tag{6}$$

where $\Delta x_i = x_j - x_j^{(*)}$, ΔR_i – small changes in the amount of allocated resources. By structure, variations Δx_j and ΔR_i, satisfying (6) do not change the value of the utility function. If the utility function is determined up to a constant multiplier with respect to money, for example, in the form of a consumer basket or other natural indicators, then the vectors p and w are proportional to the prices of services and resources. In this case, it makes sense to consider relative prices $\tilde{p}$ and $\tilde{w}$:

$$\tilde{p} = \frac{p_j}{\sum_{k=1}^{n} p_k}. \tag{7}$$

Problem (1)–(3) – parametric convex programming, which can be interpreted as an optimal production planning problem. We are not considering the problem of finding a solution to system (1)–(3), but the ability of the system to organize itself, i.e. to acquire some structure. We are interested in the mechanisms of the emergence of self-organization of the system (1)–(3) and the mechanisms of self–organization based on changes in the parameters of technologies j – parameters $\beta^{(j)}$, functions $g^{(j)}$ and coefficients of importance of services in the aggregate "consumer basket".

We will conduct a qualitative analysis of the local geometry in the vicinity of the solution of the problem (1)–(3). Let's call the optimal resource trajectories i – the trajectories of solving the problem (1)–(3) $x^{(*)}(R_i)$, resource consumption $R^{(*)}(R_i)$, prices and resource prices $w^{(*)}(R_i)$. Due to the strict convexity $g(x)$, the specific coefficients of consumption of i resources by j technologies are positive. Therefore, the function $R^{(*)}(R_i)$ changes from 0 when $R_i = 0$ before $+\infty$ by $R_i \to +\infty$. We call resource i scarce if all other resources are available in excess: $(R^{(*)}(R_i))_l < R_l,\ l \neq i$.

Resource i is called redundant if it cannot be consumed completely due to restrictions on other resources: $(R^{(*)}(R_i))_i < R_i$. In this mode, reducing the available volume of the resource does not lead to any changes. Resource i is normal if it is neither acutely deficient nor redundant. Let us consider the question of resource availability modes i. Let there be a resource R_i.

- A decrease in the volume of R_i in the mode of its acute shortage leads to: a) a monotonous drop in the volume of all types of services; b) an increase in absolute prices for resources and services; c) zero prices for other resources; d) the sameness of relative prices for services under the condition of unchanged resource availability of services j for resource i.
- In its normal mode, a decrease in R_i entails: a) a decrease in the aggregate volume of services; b) the replacement of resource-intensive services with resource-saving services; c) an increase in the absolute and relative price of a resource; d) an increase in aggregate prices; e) an increase in relative prices for resource-intensive services and a drop in relative prices for resource-saving services.

Let's check this assumption. The volume of the remaining resources is l, limited by the volumes of R_l. Let $R_i^{(\max)} = \arg\max\{R | R_i \geq 0,\ R^{(*)}(R_i) \leq R\}$. Then the increase in the volume of the resource R_i. Over $R_i^{(\max)}$ ceases to bring any useful effect to the system due to the containment of the system by the available volumes of other resources j.

An excess of resource i leads to a zero price for it. Suppose now $R_i^{(\min)} = \sup\{R_i | R_i \geq 0,\ (R^{(*)}(R_i))_i \leq R_i,\quad l \neq i\}$. That $R_i < R_i^{(\min)}$ all other l resources, $l \neq i$, are available in excess. From $R > 0$ and $R^{(*)}(0) = 0$ should $R_i^{(\min)} > 0$. Therefore, there is a regime of acute shortage of resource i, in which all other resources are available in excess and have zero price. The strict concavity of the utility function leads to an increase in absolute prices for services with a drop in R_i. Let's write out the Lagrange function of problem (1)–(3):

$$L(x, \lambda) = u(x) - (\lambda, g(x) - R), \quad \lambda \geq 0. \tag{8}$$

Maximizing it by x and minimizing it by λ, due to the excess of others other than resource i, we get: $\frac{\partial u}{\partial x_j} = \sum_i \lambda_i \partial g_i^{(j)}(x_j)/\partial x_j$, $\lambda_i > 0$. The price of service j is determined by the expression $p_j = \partial u_j/\partial x_j$, Therefore, $p_{j_1}/p_{j_2} = (\partial g_i^{(j_1)}(x)/\partial x_{j_1})/(\partial g_i^{(j_2)}(x)/\partial x_{j_2}) = const$ subject to the constant resource intensity of services j by resource i. Consider the average mode in which $R_i^{(\min)} < R_i < R_i^{(\max)}$ an increase in the volume of resource i is an additional a possibility that $R_i^{(*)} = R_i$, with strict concavity of u in x and monotonous growth and in x, leads to an increase in output. Due to the strict concavity u reduction R_i, it leads to an increase in aggregated prices for both services and resource i.

Let us now consider a small decrease in resource i in size ΔR_i, $\Delta R_i < 0$. Its aftereffect consists of the following two stages.

- At the first stage, there will be a decrease in the volume of all types of services by values $\Delta R_i / (\partial g_i(x) / \partial x)$. However, such a decrease in the volume of all types of services without exception will lead to the formation of l free, $l \neq i$ resources.
- When in normal mode, the only possibility for the system to consume l resources, $l \neq i$ is an additional reduction in the production of resource-intensive i services and their replacement with resource-saving l resource services. Thus, the second stage consists in replacing resource-intensive resource i services with resource-saving resource i services. Such substitution, due to strict concavity and monotonous growth of the utility function, leads to an increase in relative prices for resource-intensive services and a drop in relative prices for resource-saving services.

Let us now consider the change in resource consumption i by technology j: an ultra-efficient technology has negligible, compared with other normal technologies, specific coefficients of resource consumption, an extremely backward technology has immeasurably large, compared with other normal technologies, specific coefficients of resource consumption, average or normal technology has the most common specific coefficients of resource consumption. Suppose that resource i is neither redundant nor acutely deficient [3]. Then the effect of changing the resource intensity of technology j on resource i on $\Delta \partial g_i^{(j)} / \partial x_j$ consists of two stages:

- The formation of a surplus or deficit of a type i resource by an amount $x_j \Delta \partial g_i^{(j)} / \partial x_j$;
- The replacement of resource-intensive technologies of type i with resource-saving technologies of type i at $\Delta \partial g_i^{(j)} / \partial x_j > 0$ and the replacement of resource-intensive technologies of type l, $l \neq i$, with resource-saving technologies of type l, $l \neq i$, at $\Delta \partial g_i^{(j)} / \partial x_j < 0$. This is true, because the value will be $-x_j \Delta \partial g_i^{(j)} / \partial x_j$:
    - The resource i released at $\Delta \partial g_i^{(j)} / \partial x_j < 0$;
    - The resource i forming a deficit at $\Delta \partial g_i^{(j)} / \partial x_j > 0$.

By condition, resource i is neither redundant nor acutely deficient, therefore $R^{(*)}(R_i) = R$ i.e. all resources are consumed completely. If the resource intensity of service j for resource i has decreased, then the resulting excess of resource i can be consumed only by increasing the weight of technologies that are less resource-intensive for other types of resources [3, 4].

If the resource intensity of technology j for a type i resource has increased, then the resulting deficit can be covered only by replacing resource-intensive technologies with less resource-intensive ones. Now let the number of technologies be equal to the number of resource types and resource i is neither redundant nor acutely deficient.

Let's also assume that for all technologies there is a normalization condition necessary to work with relative resource intensity. Then an increase in the resource intensity of technology j for resource i leads to:

- A reduction in the volume of services of type j, if technology j is the most resource-intensive for resource i;
- An increase in the volume of services j, if technology j is the least resource-intensive for resource i.

Thus, when classifying industries by the type of resources consumed, an increase in the specific resource intensity of technology j for resource i leads to a reduction in the volume of services j if technology j is the most resource–intensive for service i, and to an increase in the volume of services j if technology j is the least resource-intensive for resource L. If the production volumes are larger than the available resources, only the general rule is true – the aggregated production of resource-intensive services will decrease, and the aggregated production of resource-saving services will grow [5, 6]. Consider the issue of direct and overhead costs. Direct expenses correspond to the terms of the first and zero orders of the function $g(x)$. Overhead costs correspond to members of the second and higher orders. Zero–order terms correspond to fixed costs, first-order terms correspond to variable costs. Members of the second and following orders may, for example, correspond to the costs of coordinating production activities. The relationship between direct and overhead costs is as follows. A decrease in direct costs – coefficients of polynomials $h^{(1)}$ and $h^{(2)}$ – on $r\%$ leads to an increase in overhead costs – coefficients of polynomials $h^{(t)}$, $t \geq 2$, on $r\%$ [7].

Based on the analysis of the optimal functioning of the company, taking into account all the specifics of choosing the most optimal management option, we will make the following recommendations:

- For each production j, there is a ratio of direct and overhead costs that minimizes resource consumption i.
- There is a ratio of direct and overhead costs for resource consumption i, which simultaneously maximizes x_j and minimizes p_j.
- If conditions a) and b) are met and technology j is the least resource-intensive for i, then:

  - Under conditions of ideal competition, there will be two types of j technologies with predominantly direct and predominantly overhead costs;
  - The goals of the manufacturer j (maximization u_j) and the goals of the system (maximization u) are inconsistent.

Now let's consider the modes of self-organization of a technical system, taking into account its features. Depending on the specific conditions, the self-organization of the system can be carried out in the following ways:

- Each of the elements maximizes its utility function $u_j(x)$;
- The system maximizes the collective utility function $u(x)$;
- The price of services is minimized – $p_j(x)$;
- The price of resources is minimized.

The first mode is achieved under the condition of independence of the actions of technology manufacturers j. The second regime can be achieved both with administrative management and with the help of management based on subsidies and taxes. The third mode is when the system is included in an ideally competitive environment, which, due to competition, will minimize the prices of services. The fourth mode is with an ideal resource market: the greater the difference between the service and the unit cost of the resource, the higher the company's profit.

Consider the following version of the problem (1)–(3):

$$u(x) = \sum_{j=1}^{n} \alpha_j \ln(x_j) \rightarrow \max, \tag{9}$$

$$g_i(x) = \sum_{j=1}^{n} (C_{i0j} + C_{i1j}x_j + C_{i2j}x_j^2) \leq R_i, \tag{10}$$

$$x \geq 0, \tag{11}$$

where $\alpha_j > 0,\ C_{i0j} \geq 0,\ C_{i1j}x_j \geq 0,\ C_{i2j} \geq 0.$

That is, we refine the problem (1)–(3), use the utility function $u(x)$ according to the Cobb-Douglas type [6].

The coefficients C_{i0j} in Eqs. (10) are not necessarily zero. For example, the company would consume electricity, heat even with zero output. Thus, the coefficients C_{i0j} reflect the fixed costs of resource i for service j.

Consider the possibility of an analytical solution to our problem, for this we assume $m = 1$ and $\mathrm{C}_{i0j} = 0$ for all j. Then problem (9)–(10) has an analytical solution:

$$x_j = \{\alpha_j \alpha_0^{-1} (R - C_0)/C_{2j}\}^{1/2}, \tag{12}$$

$$p_j = \{\alpha_j \alpha_0^{-1} C_{2j}/(R - C_0)\}^{1/2}, \tag{13}$$

$$w = \alpha_0/(2(R - C_0)), \tag{14}$$

where $\alpha = \sum_{i=1}^{n} \alpha_j,\ C_0 = \sum_{j=1}^{n} C_{10j},\ C_{2j} = C_{12j},\ R = R_1.$

It follows from the analytical solution (12)–(14) that with an increase in the reserves of a scarce resource (R):

- All issues of x_j increase, although to varying degrees;
- The price of p_j services and the price of the resource w decreases;
- The "specific utility" of a resource unit decreases, which can be expressed by the formula $u(R)/R$.

It also follows from the same formulas that with an increase in a certain coefficient of additional costs, the volume of services with a number $j(x_j)$ decreases, and its price increases.

At the same time, the volume and prices of other services, as well as the price of the resource remain unchanged. When changing the amount of available resources *R*, the following results are obtained. With small values of the resource being changed, the remaining resources are in excess. There are three main modes of the system.

1. The replaceable resource is in excess.
2. The replaceable resource is acutely deficient.
3. The resource being exchanged is neither excessive nor acutely deficient.

For the first and second modes, it is possible to stabilize relative prices for resources and services.

For the third mode, the mode of transition from the second state to the first, with an increase in the volume of the resource being exchanged, the price of the resource being exchanged decreases linearly with an increase in the volume of the resource being exchanged, for example, with an increase in the volume of the resource R_1. In this situation, there is a change in prices for services, and as the volume of the resource being exchanged decreases, the price for services of the most resource-intensive technologies increases and decreases for services that are least resource-intensive. This is due to the displacement of resource-intensive services by resource-saving ones. As a result, the supply of resource-intensive services is reduced by α and the supply of resource-saving services is growing. The curtailment of supply leads to an increase in price, an increase in supply leads to a decrease in price. In the first mode, the mode of an obvious shortage of a replaceable resource, the volume of all services is curtailed as the volume of an acute resource falls [8, 9].

3 Result and Conclusions

The collective utility function monotonically increases with an increase in the volume of the resource being exchanged, reaching its maximum at the first stage. Indeed, at the first stage – the stage of an excess of a replaceable resource, its further increase does not affect the volume of services provided, therefore, the collective utility function. The specific collective utility function per unit of the resource used $-u/R_i$ – reaches its maximum at a certain value R_i. The maximum is not achieved in saturation mode.

The practical use of such an approach to optimizing management decisions in a company based on functional and methodological modeling allows us to conclude that there are three modes of the system's response to changes in C - the introduction of resource–saving technologies depending on three possible states of product production technology:

- Ultra-efficient technology;
- Extremely backward technology;
- Medium or normal technology.

In the first mode – the mode of ultra-efficient technology – a decrease in the coefficient C leads to a decrease in the volume of super-resource-saving services. The reason is the need to save other resources, the consumption of which other technologies are more resource-saving.

In the second mode, the mode of extremely backward technology, there is a de facto leaching of technologically backward services from production – the service is replaced by less resource-intensive services. A decrease in resource intensity – the coefficient C – leads to an increase in the volume of services provided. At the same time, the price of an excessively consumed resource becomes dominant, exceeding the prices of other resources by an order of magnitude. The price of excessively resource-intensive services also becomes dominant, exceeding the prices of other services by orders of magnitude. This is happening against the background of a sharp decline in the output of resource-intensive services. The second mode is observed, for example, when responding to one or more technological revolutions.

In the third mode – the mode of transition from the first state to the second state, for example, with a gradual decrease in resource intensity – a decrease in C can lead to both an increase and a decrease in the volume of services, depending on the relative scarcity of other resources and the characteristics of others technologies.

References

1. Ayvazyan, S.A.: Intellectualized instrumental systems in statistics and their role in the construction of problem-oriented decision support systems. Rev. Appl. Industr. Math. **4**, 79–85 (1997)
2. Baev, L.A.: Intensive Self-organization of Economic Systems. Concept, Theory, Models, 268 p. CHSTU, Chelyabinsk (1992)
3. Bazilevich, L.A.: Modeling of Organizational Structures, 159 p. LSU Publishing House (1978)
4. Sharakshane, A.S., Zheleznov, I.G., Ivanitsky, V.A.: Complex Systems, 110 p.Higher School, Moscow (2017)
5. Fedorenko, N.L.: Economic and Mathematical Models in the Enterprise Management System, 174 p. Nauka, Moscow (1998)
6. Abrial J-R. Data semantics. In: Klimbie, J.W., Koffeman, K.L. (eds.) Data Management Systems, North-Holland, pp. 78–92 (1974)
7. Bubenko, J.A.: Information Systms Methodologies. North-Holland, Amsterdam, The Netherlands, pp. 289-318 (1986)
8. Klein, H.: Towards a new understanding of data modeling. In: Floyd, C., Züllighoven, H., Budde, R., Keil-Slawik, R. (eds.) Software Development and Reality Construction, pp. 203–220. Springer, Heidelberg (1992). https://doi.org/10.1007/978-3-642-76817-0_17
9. Lyytinen, K.: Penetration of information technology in organizations. Scand. J. Inf. Syst. **3**, 41–68 (1991)

Overview of the DJI ZENMUSE L1 Laser Scanner for Spatial Planning Purposes

Dmitry Gura[1(✉)], Elizaveta Berkova[1], Anastasia Panyutischeva[1], Monji Mohamed Zaidi[2,3], and Gennadiy Turk[1,4]

[1] Kuban State Technological University, 2, St. Moskovskaya, Krasnodar 350072, Russia
gda-kuban@mail.ru

[2] Department of Electrical Engineering, College of Engineering, King Khalid University, Abha, Saudi Arabia

[3] Electronics and Microelectronics Laboratory (LR99ES30)-FSM, Monastir University, Monastir, Tunisia

[4] Kuban State Agrarian University named after I.T. Trubilina, 13, St. Kalinina, Krasnodar 350044, Russia

Abstract. Laser scanning of objects is a modern method for obtaining 2D and 3D models of the surrounding space. During the operation of the devices, a cloud of points with spatial coordinates is created, which ultimately give a three-dimensional image. Three-dimensional scanning of objects allows you to create digital models not only of individual buildings and structures, but also of entire complexes or territories. Laser scanning technology has existed for a long time, the effect of this technology was felt by specialists in various industries only in the last decade. In geodesy, laser scanning serves as a source of data for creating accurate three-dimensional models. By using drones to quickly collect data from a safe distance, it becomes possible to instantly receive up-to-date information that does not require extra costs. LiDAR technology was introduced into the DJI Zenmuse L1 system, thereby creating a payload module that can translate point clouds in real time and collect data to develop a topographic map of the area. The technology for processing laser scanning data is considered using the DJI ZENMUSE L1 laser scanner as an example. Methods and technologies for processing scanning data at all stages have been collected and combined - from the preliminary processing of "raw data" to the creation of digital elevation models, contour lines and other products. The functionality of DJI Terra and Spatix/Terrascan programs is described. The aim of the study is to study the principles of laser scanning data processing. The research problem is the imperfection of laser scanning data processing methods. As a result of the work done, using the DJI Terra program, the data obtained using the 3D airborne laser scanning technology was processed. With active LiDAR detection, Zenmuse L1 can scan terrain to collect point clouds even at night or in low light conditions with an integrated RGB camera and a high-precision IMU.

Keywords: Laser Scanning · Point Clouds · Data Processing · Reflection · Lidar · Sustainability Goals

© The Author(s), under exclusive license to Springer Nature Switzerland AG 2023
A. Gibadullin (Ed.): DITEM 2022, LNNS 683, pp. 32–41, 2023.
https://doi.org/10.1007/978-3-031-30926-7_4

1 Introduction

1.1 General Principles of Laser Scanning

Laser scanning is a survey method that helps to quickly acquire the most complete and reliable spatial and geometric information about the objects of the study area. The technology of three-dimensional laser scanning is based on the method of determining the set of three-dimensional coordinates X, Y, Z of individual points on the object being photographed. And the principle of laser scanning itself is quite simple: some equipment emits a laser beam and then another part of this equipment, which is called a receiver, receives this beam, and then the program calculates the distance to the point from which this laser beam was reflected. In this case, it is necessary to know the additional parameters of the GCC (an optimizing compiler created by the GNU project that supports various programming languages, hardware architectures and operating systems), the inertial component in order to subsequently determine the high-precision coordinates of the points from which the reflection was received. In the simplest theoretical sense, laser scanning is the determination of the distance to an object using a laser beam [8, 12].

Laser scanning is performed using lidar. The definition of "lidar" is an abbreviation for the full name "lightdetectionandranging", which means detecting and ranging with the help of light.

DJI Zenmuse L1 is a high-tech device in topographic mapping due to the fact that this system, as a result of the work carried out, can create high-precision digital terrain models, digital elevation models, solve problems of territorial planning and monitoring, and also perform many other specialized projects.

The Zenmuse L1 system combines the most advanced and best technologies: LivoxLidar module, high-precision IMU module and a powerful 1-inch CMOS camera. And it all works together with DJI's 3-axis stabilization system. To get a complete solution for displaying 3D data in real time, it is necessary to combine an industrial drone, the DJI Terra software application and the L1 system into a single complex.

With the ability to actively detect the LiDAR module, the Zenmuse L1 can scan the terrain to collect point clouds even at night or in low light conditions. The module is a robust design with an IP44 protection class, which means that it is not afraid of precipitation and bad weather conditions. With an integrated RGB camera and a high-precision IMU, virtually any task assigned to the system will be completed.

1.2 Purpose of the Study

This article discusses the processing of laser scan data using DJI Terra and Spatix/Terrascan software. The work of the DJI Zenmuse L1 laser scanner was also considered, the purpose of which is the development of 3D models of the territory with a high degree of detail, as well as flat drawings and sections [4, 20].

2 Materials and Research Methods

2.1 DJI L1 Laser Scanner Features

DJI L1 features include the presence of the LiDARLivox module, which supports up to three displays, and also captures 240,000 points per second with a single reflection, the scanning angle of which is 70°. It works in two scan options:

- Non-repeating lobe sweep, the viewing angles of which are 70.4° × 77.2°. This method is better suited for tasks related to determining the characteristics of the forest, as well as for performing specific work, since there are several measurements at different angles for the same area of the terrain;
- Line scan. The viewing angles are 70.4° × 4.5°. This option is less noisy and better suited for obtaining a digital model.

The DJI L1 Scanner has 3 reflection capture options:

- Fixes the last reflection;
- Fixes the first and last reflection;
- Fixes the first, last and center reflection.

Reflectivity refers to the ratio of reflected laser energy to received laser energy; surfaces of different objects have different reflectivity (Table 1).

Table 1. Reflectivity of surfaces.

Surface type	Reflectivity
Sand	0.24–0.28
Earth	0.1–0.2
Wet surface	0.08–0.08
Dry grass	0.15–0.25
Wet grass	0.14–0.26
Snow	0.81

The reflectivity on most soil surfaces exceeds 10%, on the surface of buildings - about 50%, on asphalt - up to 20% [14, 15].

The DJI L1 is the lightest laser scanner with an integrated camera. It has LiDAR and an RGB camera. This lidar has the following functionality:

- Vertical accuracy - 5 cm, and horizontal - 10 cm;
- Shooting efficiency - 2 square kilometers per flight;
- Measurement speed - up to 480,000 points per second;
- Protection against dust and moisture - IPS4;
- The ability to display a cloud of points in real time;
- Possibility of post-processing in DJI Terra.

2.2 Operating the DJI L1 Laser Scanner

To work with lidar, additional equipment is also required. For example, a three-axis stabilized gimbal, which allows you to keep the lidar in the right direction. It provides stable and uniform coverage of the territory by scanning and photographic data [9, 11].

Zenmuse L1 supports broadcasting a point cloud in flight mode to the remote control, that is, information from the lidar will be displayed on the remote control. This allows you to monitor the progress of work and identify errors and discrepancies. When such errors are detected, it becomes possible to rebuild the shooting parameters so that there are no more discrepancies, or create a new mission. The operating efficiency of the DJI L1 is as follows:

- Flight altitude - 100 m;
- Transverse overlap - 20%;
- Flight speed - 10 m/s;
- Flight time is about 30 min [16, 19].

Displaying point cloud data in the DJI app in real time for monitoring the progress of work and making emergency decisions is used in the case of:

- Displaying the model in real time when scanning L1;
- Switching between the point cloud in real time and the view from the camera;
- Using multiple color modes: reflectance, height, RGB mode.

Post-processing takes about half of the flight time. The point clouds are displayed in real time. After enabling the IMU calibration option, calibration zones are automatically added at the beginning, end and in the middle of the route (if the route is longer than 100 s). Calibration is a single acceleration-deceleration action on a 30 m segment. Calibration is a single acceleration-deceleration action on a 30 m segment. If the route is very long, for example more than 200 s, then additional calibration zones are added, one for each 100 s [18].

Data Processing. Lidar supports two processing methods: RTK and PPK, which are processed in DJI Terra Pro. The sequence of work is shown in Fig. 1:

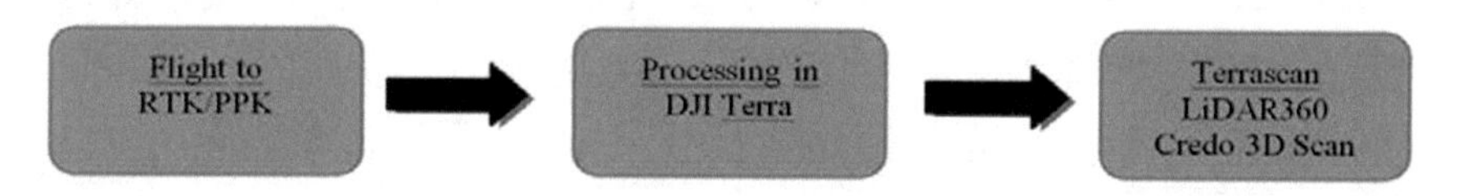

Fig. 1. Sequence of data digitization.

For PPK using its own database, it is necessary to copy it to the "Measurement file" folder and rename it similarly to system files. After the flight is completed, the following file directory is created:

CLC is Lidar and camera calibration file, that is, these are the files that are necessary in order to bring all measurements to one conditional point;

CLI is lidar and inertial navigation systems (INS) calibration file;

CMI is visual sensor calibration file;

IMU is ANN data, which constantly records accelerometer and gyroscope data;

LDR is lidar data, which can be a number of files;

RTB is a file of measurements at the base station when flying in RTK, that is, the system records the necessary data in order to process them later;

RTK is global navigation satellite system (GNSS) binary file of main antenna observations;

RTL is technical data (size of guides, etc.);

RTS is a binary file of GNSS observations of the second antenna, that is, it is additionally involved in the processing process, including to improve the accuracy of heading;

EVENTLOG.bin is a binary file with camera timestamps. In general, this is a single standard format for storing timestamps for pictures and other additional information used by DJI;

PPKRAW.bin is binary file of GNSS observations for images;

MRK is file with timestamps for snapshots;

JPG is pictures during the flight.

When launching DJI Terra, a Google maps appears, this can be changed to a satellite maps. It displays the flight restriction zones that operate on the territory of the country. To process laser scan data, you need to create a new job. The program gives you a choice of 3 types of tasks:

- Visible light;
- Multi spectrum;
- Lidar point cloud [1].

In order to carry out the processing of laser scanning data, a third type is required: "lidar point cloud". In this window it is possible to change various parameters. It is necessary to set the settings for the center point of the base station. In Russia, the coordinate system (CS) SK-42 or SK-95 is used, but it is usually performed in WGS 84 CS, since all satellite measurements are processed in this CS. All ellipsoids and ground parameters have transitions to WGS 84. In the center point settings, it is possible to change the following parameters:

- Lidar point cloud. The parameter allows you to select the required density for work;
- Scripts. In this window, the point cloud is processed, the Zenmuse L1 is calibrated;
- Advanced. Here are the most options, such as:

  - Effective distance of the point cloud. The option makes it possible to optimize the accuracy of the point cloud;
  - Coordinate system at the output. In this window, the SC is selected (conditional and specified). It is necessary to select a given SC, however, the default is the SC

in which the flight was made. Also here is the geoid setting. There are quite a lot of geoid parameters, but in Russia, EGM2008 is mainly used;

- Altitude offset is the total displacement of the cloud by a given value;
- Output format, default is PNTS;
- Consolidated output. In this window, you can choose how the data will be processed: separately or will be merged into a single point cloud;

– Applications, annotation and measurements.

3 Results of the Research and Their Discussion

When you have finished changing the settings, the Reconstruction Parameters Checklist window appears. Here you can check if the settings are correct. The data processing begins. After loading the data, the system makes it possible to select the option for displaying the point cloud: RGB, reflectivity, reflected wave height. With these data, you need to go to the "Spatix" program. "Spatix" is a point cloud visualization platform. This platform has tools related to vector work. It can be used to process laser scan data. On Fig. 2 shows the "Terrascan" dialog box [2].

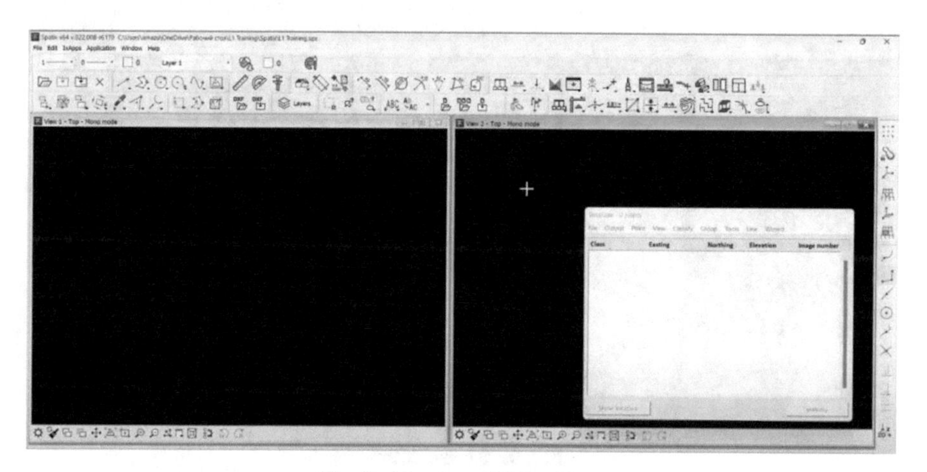

Fig. 2. Terrascan Dialog Box.

The "Wizard" assistant allows you to process laser scanning data. It performs a number of functions:

- Ability to view different trajectories of the scanner;
- Cutting out overlaps, since this data is not needed when using lidar, as well as removing extra points;
- View spans (Fig. 3);
- Smoothing and noise removal;
- Classification of cloud points according to certain parameters (Fig. 4).

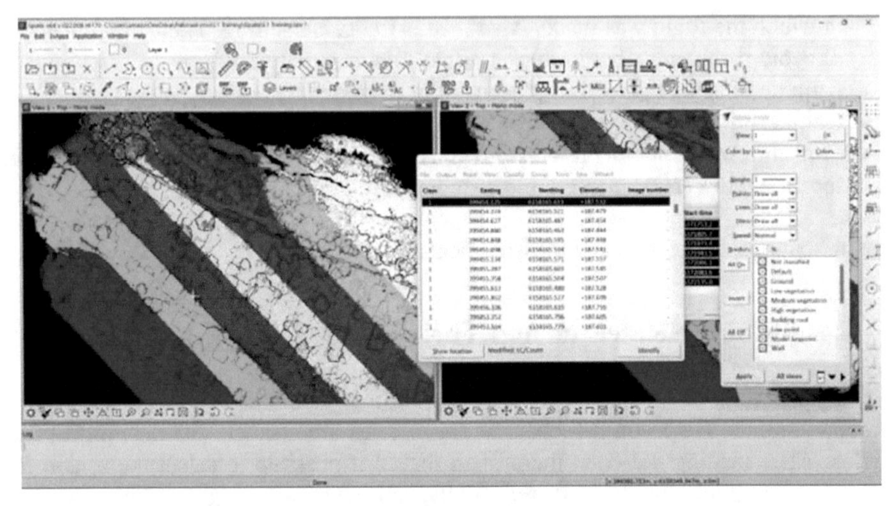

Fig. 3. View spans.

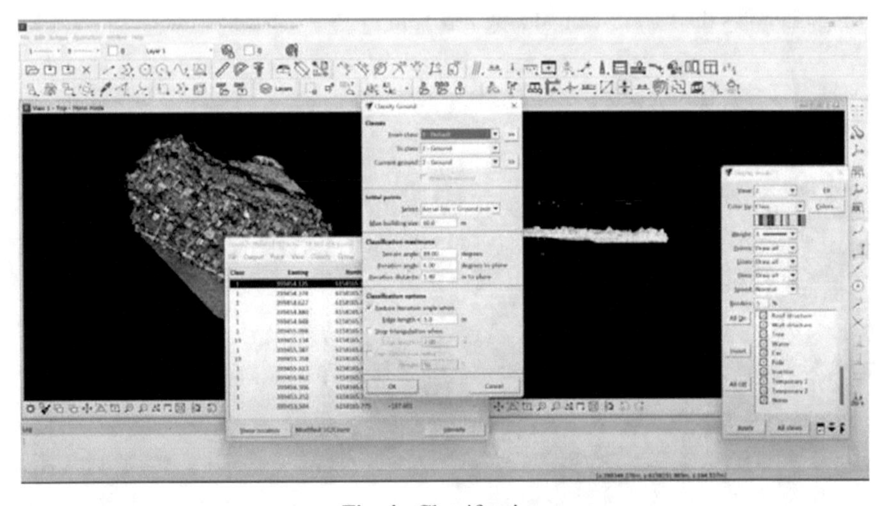

Fig. 4. Classification.

In addition to the functions listed above, the program provides the creation of contour lines, as well as the classification of various types of objects. The classification result is shown in Fig. 5.

The program makes it possible to classify any objects. For example, each tree can be separately measured and placed in separate groups (Fig. 6). You can get the calculation of the coordinates of the middle of the trunk [13, 17].

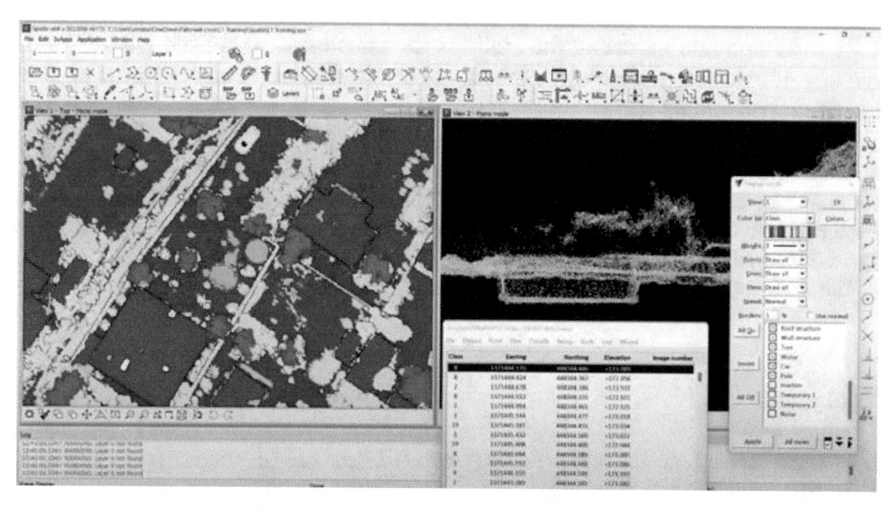

Fig. 5. The result of object classification.

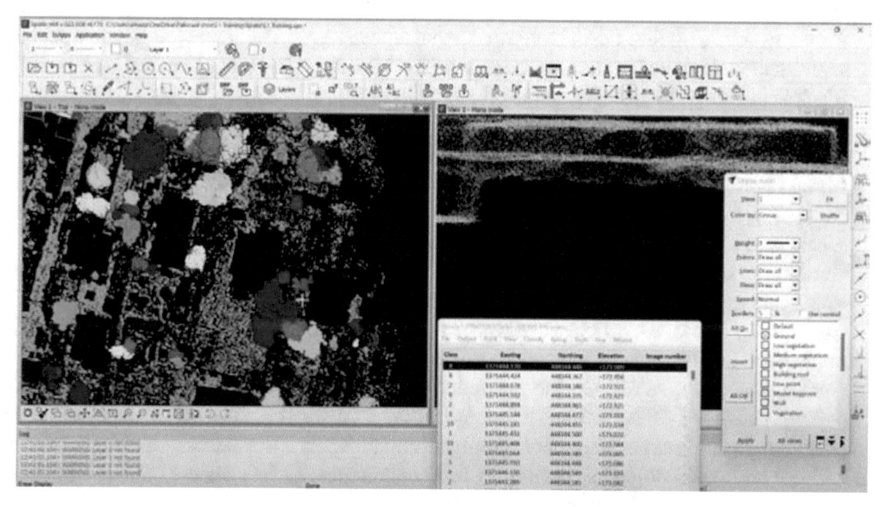

Fig. 6. Classification of trees.

4 Conclusion

Thus, the DJI Zenmuse L1 laser scanner is the most convenient device for obtaining information about the terrain, since it belongs to the lightest lidars with an integrated camera [3]. It is highly likely that the work with this lidar will be the most effective in comparison with other analogs [10, 18].

As a result of the work done, using the DJI Terra program, the data obtained using the 3D laser scanning technology was processed [5]. The program allowed to adjust the parameters in accordance with the required result of the work performed, and also provided many functions that allow you to process all the information that was received

[6, 7]. An important aspect of the study is the conclusion about the possibility and necessity of using the technology of three-dimensional laser scanning for the purposes of territorial planning.

Acknowledgments. The study was carried out using the equipment of the Research Center for Food and Chemical Technologies of KubGTU (CKP_3111), the development of which is supported by the Ministry of Science and Higher Education of the Russian Federation (Agreement No. 075-15-2021-679).

The study was supported by the Russian Science Foundation grant No. 22-29-00849.

References

1. Webinar “Laser scanning data processing with DJI Zenmuse L1”. https://youtu.be/K3yYBzH5HUU. Accessed 10 Nov 2022
2. Altyntsev, M.A.: Linking mobile laser scanning data to aerial photography results based on determining the relative position of arrays of points. Bull. SGUGiT (Siber. State Univ. Geosyst. Technol. 5–15 (2022)
3. Arakelov, M.S., Lipilin, D.A., Dolgova-Shkhalakhova, A.V.: The impact of quarantine measures against the new coronavirus infection COVID-19 on the state of coastal waters of the Black Sea. Geogr. Environ. Sustain. **14**, 199–204 (2021)
4. Gorkavy, I.N.: Comparative analysis of methods for obtaining a model of the Earth’s surface from three-dimensional laser scanning data. News of higher educational institutions. Geodesy Aerial Photogr. 58–62 (2011)
5. Karpanina, E.N., Gura, A., Ron, I.N.: Rationale of the system approach to education of future specialists in the university. Astra Salvensis **6**, 763–765 (2018)
6. Gura, D.A., Dyachenko, R.A., Khusht, N.I.: On the issue of preprocessing three-dimensional laser scanning data. Electron. Netw. Polythemat. J. “Sci. Works KubSTU” 39–46 (2021)
7. Gura, D.A., Markovsky, I.G., Ryaskin, A.A.: The use of unmanned aerial vehicles in the implementation of state land supervision. Bull. SGUGiT 138–146 (2022)
8. Gura, D.A., Gura, A.Yu.: On the methodology of training university students in the competence geodesy G60 according to the standards of “Worldskills Russia”. Quality of higher education in an agrarian university: problems and prospects: a collection of articles on educational and methodological materials of the conference. Kuban State Agrarian University named after I.T. Trubilin, pp. 93–96 (2019)
9. Freydin, A.Ya.: Three-dimensional laser scanner: principle of operation and scope of application. World Meas. 47–49 (2007)
10. Kosolapov, P.A., Dyachenko, R.A., Gura, D.A., Khusht, N.I.: On the normalization of three-dimensional laser scanning data. Electron. Netw. Polythemat. J. “Sci. Works KubSTU” 56–66 (2021)
11. Medel, F., Abad, J.S., Esteban, V.: On the use of laser-scanning vibrometry for mechanical performance evaluation of 3D printed specimens. Mater. Design (2021)
12. Sarksyan, L.D., Lukyanova, M.S., Solodunov, A.A., Pshidatok, S.K.: Types of laser scanning and their features, pp. 83–86 (2019)
13. Sithole, J.: Filtering of laser altimetry data using an adaptive tilt filter. In: International Archive of the Sciences of Photogrammetry, Remote Sensing and Spatial Information. Annapolis, Maryland, pp. 22–24 (2001)

14. Shevchenko, G.G., Gura, D.A., Gura, A.Yu., Chernova, N.V.: Database “three-dimensional coordinates of marks of a multi-storey residential building for determining displacements and precipitation”. Certificate of registration of the database RU 2018621135, 24.07.2018. Application, No. 2018620403 (2018)
15. Schnell, G., Duenow, U., Seitz, H.: Effect of laser pulse overlap and scanning line overlap on femtosecond laser-structured ti6al4v surfaces. Materials 969 (2020)
16. Sukhomlin, V.A., Bitter, I.N.: Technological system for building software complexes for automation of processing and 3D laser scanning data. Comput. Sci. Appl. 53–64 (2009)
17. Voselman, G.: Filtering of laser altimetry data based on tilt. In: International Archives of Photogrammetry, Remote Sensing and Spatial Information Sciences, Amsterdam, pp. 23–27 (2000)
18. Wang, Yu., Mercer, B., Tao, V.S., Sharma, J., Crawford, S.: Automatic generation of digital models of bare earth relief from digital surface models created using on-board. Ifsar, Continuation of ASPRS, pp. 935–942 (2001)
19. Yang, L., et al.: Automatic guidance method for laser tracker based on rotary-laser scanning angle measurement. Sensors 1–17 (2020)
20. Brouček, J.: Comparing mobile laser scanning and static terestrial laser scanning. In: 19th International Multidisciplinary Scientific Geoconference SGEM, Sophia, pp. 83–90 (2019)

Forecasting Uneven Time Series: From Accuracy to Efficiency

Oleg Russkov, Sergei Saradgishvili, Nikita Voinov(✉), and Anton Tyshkevich

Peter the Great St. Petersburg Polytechnic University, 29, Polytechnicheskaya, St. Petersburg 195251, Russia
voinov@ics2.ecd.spbstu.ru

Abstract. The problems of using existing models in forecasting uneven time series are described. The application of game theory in uneven power consumption forecast models is substantiated. The complex issue of the transition from a simple forecast error reduction to effectiveness of forecast is considered. The adjustment of the model parameters to increase the forecast effect is shown. The conclusion about the applicability of the described algorithms is made.

Keywords: Power Consumption · Uneven Time Series · Forecasting · Game Theory

1 Introduction

The work of most modern time series forecasting models is estimated by forecast error. The most efficient model has the smallest forecast error. Rapid development of digital technologies in the wholesale electricity and gas markets leads to the widespread introduction of hourly planning the energy consumption. It increases the role of forecast accuracy [1]. Forecasting hourly energy prices is also important [2]. The most popular forecasting models in the world [3] are neural network [4] and autoregressive ones [5] now. Although some researchers that image recognition is the main field of neural networks application. Autoregressive models have disadvantages too [5]. Nevertheless, using the models described above is quite justified for relatively uniform time series. The forecasting error and the speed of these models completely suit their users [3]. However, increasing the unevenness or volatility of time series leads to increasing the forecast error. It creates difficulties of applying these models in practice [6, 7]. Examples of uneven time series that cause similar forecasting problems are energy consumption series [8] or stock exchange price indexes during periods of volatility [9]. Therefore the development of methods and models for forecasting uneven time series is important scientific, technical, economic and production task [10]. The authors of this article develop the method for forecasting uneven power consumption previously described in the scientific literature [6–8, 10].

© The Author(s), under exclusive license to Springer Nature Switzerland AG 2023
A. Gibadullin (Ed.): DITEM 2022, LNNS 683, pp. 42–51, 2023.
https://doi.org/10.1007/978-3-031-30926-7_5

2 Materials and Methods

2.1 The Method for Forecasting Hourly Power Consumption Based on Game Theory

The most common and scientifically significant time series in the energy sector are time series associated with generation and consumption of electricity [11]. It is justified by the requirements of hourly accounting and full digitalization of wholesale electricity markets all over the world. The time series of industrial enterprises power consumption is closely related both to the series of electricity generation and to the series of hourly prices of its purchase and sale. Therefore in order to solve the problem of reducing time series forecasting error it is necessary to apply a systematic approach, because of the complexity of using existing forecast models [6, 7]. Related series to power consumption of industrial enterprise are hourly electricity prices. Time series of electricity generation have an impact on consumption series but are not related. Having identified the presence of adjacent time series it is necessary to analyze the possibility of their forecasting and further formalization of their connection with the time series of electricity consumption. Therefore the ratios of two parameters of each of the adjacent series are the subject to analysis. First parameter is the forecast value, second - the factual one. Formation of each of these parameters is a non-deterministic experiment. Only one of two mutually exclusive outcomes can be realized (Fig. 1).

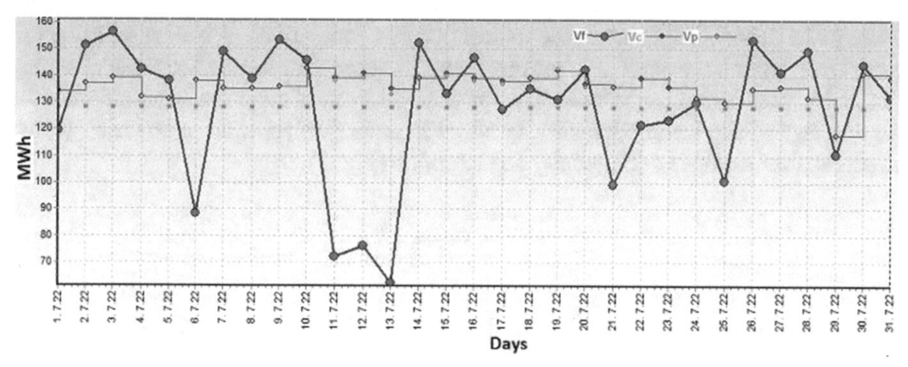

Fig. 1. Factual Vf, planned Vp and corrected Vc values of industrial enterprise power consumption.

The case when the forecast value coincides with the factual one is an exception to this rule. Figure 1 shows the statistics of factual V_f, planned V_p and corrected V_c values of industrial enterprise power consumption per hour 14–15 for 1 month. Figure 2 shows the statistics of the Day ahead market hourly prices DAM_{buy} and Balancing market prices BR_{buy}, BR_{sell} of adjacent time series per hour 14–15 for 1 month. The statistics is provided by software forecast model "Flow3" based on described approach.

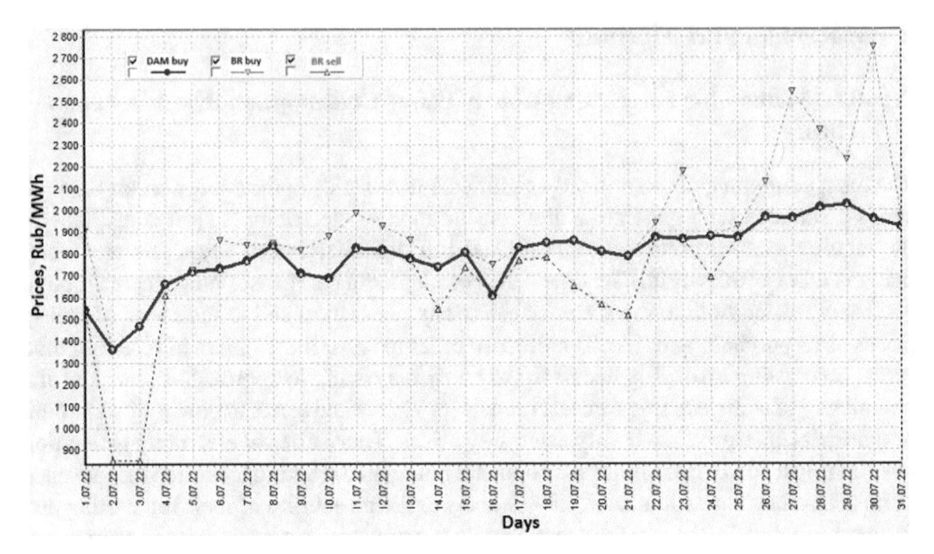

Fig. 2. Prices DAMbuy, BRbuy, BRsell of industrial enterprise per July 2022.

For a formal description of the process of implementing the outcomes of these experiments, it is permissible to use the tools of probability theory [12]. The market participants compete with each other at the day ahead auction every hour. Therefore using game theory [13] is advisable to describe this competition. An electronic auction every hour generates a conflict of players interests, since their forward-looking bids may not be accepted by energy market. The forecasted values of electricity consumption for the next day within the auction directly affect the values of adjacent price series, which affect the future values of electricity consumption (Fig. 3).

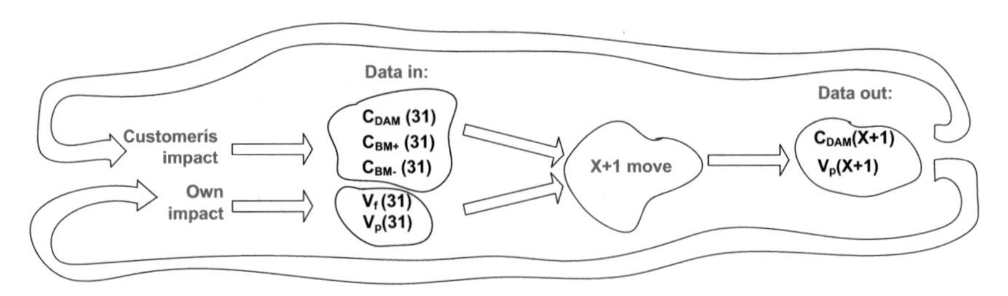

Fig. 3. Market players game.

The game of the market participants is a sequential non-cooperative game with a non-zero sum and incomplete information. In order to minimize the uneven time series forecast error it is necessary to solve the problem of choosing the magnitude and direction of preliminary forecast correction from a variety of alternatives (Fig. 4).

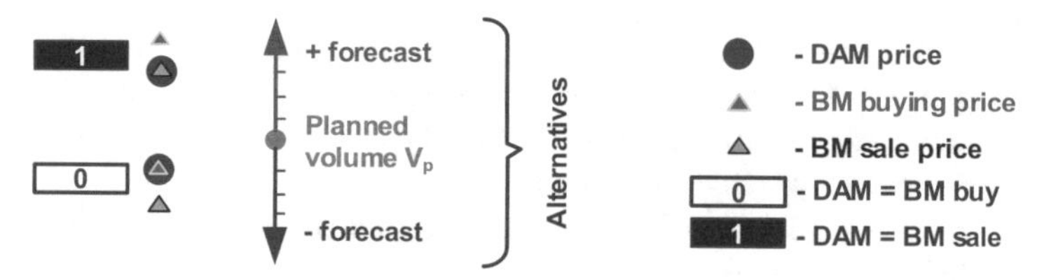

Fig. 4. Variety of forecast alternatives.

2.2 Transition from Error to Efficiency of Forecasting

The target function of forecasting uneven time series generally contains two criteria [8], however, the single-criterion problem is solved much easier. Therefore in the case of forecasting uneven time series through an adjacent one, the forecast error can be expressed in their cost. It leads the decision-making task to single-criteria one [8]:

$$\left\{\sum_{i=1}^{N} (S^{i}_{DAM+BM} + C_{dif} \cdot |V_i|) < \sum_{i=1}^{N} (S^{*i}_{DAM+BM} + C_{dif} \cdot |V_i^{*}|)\right. \tag{1}$$

where S^{i}_{DAM+BM} is the cost of hourly power consumption and V_i is the amount of deviations of factual value from forecasted one after correcting the forecast of uneven series, $S^{*}_{DAM+BM}i$ is the cost and V_i^{*} is the amount of deviations before correcting, i is the hour number, N is the number of hours in the reporting period.

The described relationship between the adjacent and forecasted time series makes it possible to move from the concept of forecasting uneven power consumption error to the effectiveness of joint forecasting all adjacent time series. At the same time, according to the obtained target function, the loss in the forecasting main series error can be compensated by more accurate forecasting the adjacent series. This means reaching a more complex characteristic – forecasting efficiency. It more accurately describes the processes taking place in the wholesale electricity market. The forecast model based on this approach compares the probability of obtaining a benefit and risk of loss with different alternatives to the forecast values of time series.

2.3 Setting up the Parameters of Forecast Model

It is possible to adjust the parameters of the predictive model during experiments. The main parameter is the duration of the statistical data sampling. At the same time the authors have experimentally proved that increasing the number of days does not lead to decreasing the forecast error - the necessary and sufficient duration of period is 31 days [8]. This duration takes into account seasonality and corresponds to the size of reporting period in energy sector. Next important parameter of forecast model is the ratio of positive and negative parts of the target function calculated using the full probability formula for all possible ratios of the forecasted and adjacent time series [8]. The positive part of the target function responsible for the positive correction characterizes the achievement of profit:

$$F_{1+} = V \cdot (P_1 \cdot \Delta_0 + P_1 \cdot P_2 \cdot \Delta_1 + C_{dif}) \tag{2}$$

where P_1 is the probability of forming type 0 price ratio; V - the magnitude of the forecast value correction; P_2 - the probability that factual value will be greater than forecast one; Δ_0, Δ_1 - the average price difference for ratios of types 0 and 1, respectively; $C_{dif.}$ - the price of deviations in the analyzed interval [8].

The negative part of target function characterizes the receipt of a possible loss respectively [8]:

$$F_{1-} = -V \cdot (P_2 \cdot \Delta_1 + P_1 \cdot P_2 \cdot \Delta_0 + 2 \cdot P_2 \cdot C_{dif}) \tag{3}$$

The correction is made depending on the attitude F_{1+} to F_{1-}. Experiments show that the highest value of the target function corresponding to $F_{1+}/F_{1-} = 1,1$ (Fig. 5). The sign of correction depends on the prevalence of corresponding target function (F_0 for negative correction is opposite in sign F_1 for positive one). The value of corrected forecast value V_c (Fig. 1) corresponds to the maximum economic effect obtained as a result of forecasting uneven power consumption.

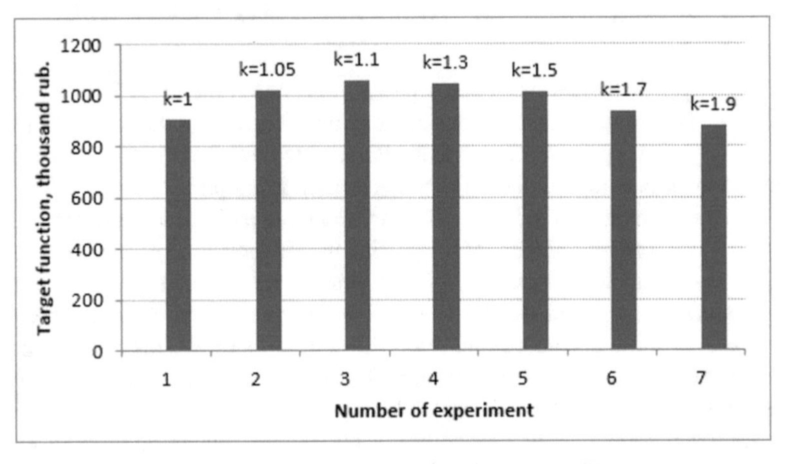

Fig. 5. Depending the target function on F_{1+}/F_{1-}.

Thus the algorithm for applying the target function is formalized. It takes into account the effectiveness of forecasting two adjacent time series, and not only the error of forecasting uneven main series. This approach allows to bypass the limitations of existing forecast models.

3 Results and Discussion

The forecast model is implemented as software package in C++ language. Experiments are carried out in retro forecast mode. Submission of bids for the day ahead is made in online forecast mode. The software is used in the department of chief power engineer of large metallurgical enterprise.

The transition from error to efficiency is illustrated as follows. The application of described forecast model in 2021 showed increasing the error of forecasting electricity consumption by 0.3% relative to "average" forecast. However, at the same time, due to forecasting prices relations the model achieved an economic benefit in amount of 1.377 million rubles at day ahead market (Fig. 6).

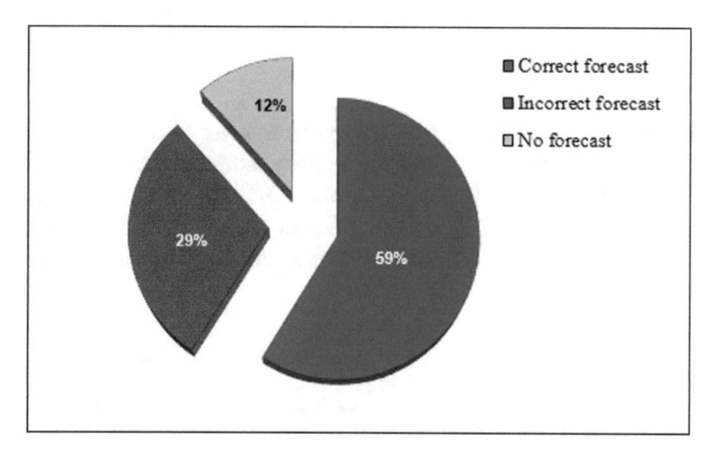

Fig. 6. Hourly prices relations forecast for industrial enterprise per 2021 year.

In addition, autocorrections of prices relations forecast error [8] was recorded at the red zone of Fig. 6 (14%). In 12% of cases the forecast, taking into account the statistics of time series relations, was not fulfilled. As a result the final effect of the forecast is positive and amounted to 0.763 million rubles despite the loss from increasing the electricity consumption relative forecast error by 0.3% or 0.615 million rubles. It proves that following the current trends of adjacent time series to correct the forecast of main time series makes it possible to increase the effectiveness of joint forecast even despite the growth of main series forecast error (Fig. 7).

Therefore in the case of forecasting uneven time series it is logical to use the concept of "forecasting efficiency" instead of "forecasting error". This makes it possible to more accurately reflect a comprehensive approach to joint forecasting of uneven and adjacent time series.

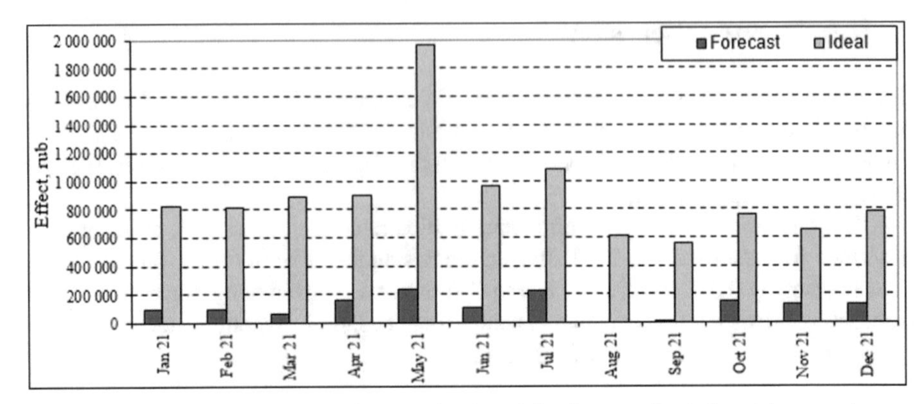

Fig. 7. The effect of electricity volume and prices joint forecast for industrial enterprise per 2021 year.

If forecast error of main time series decreases as error of adjacent series, then the effect of joint forecast significantly increases. For example the effect of July 28, 2022 prices relations forecast was 7.5 thousand rubles. The error of electricity consumption forecast was reduced by 1.3%. The overall effect increased by 2 times to 15.1 thousand rubles. The distribution of 13 previously described cases [8] of joint forecast is shown at Fig. 8.

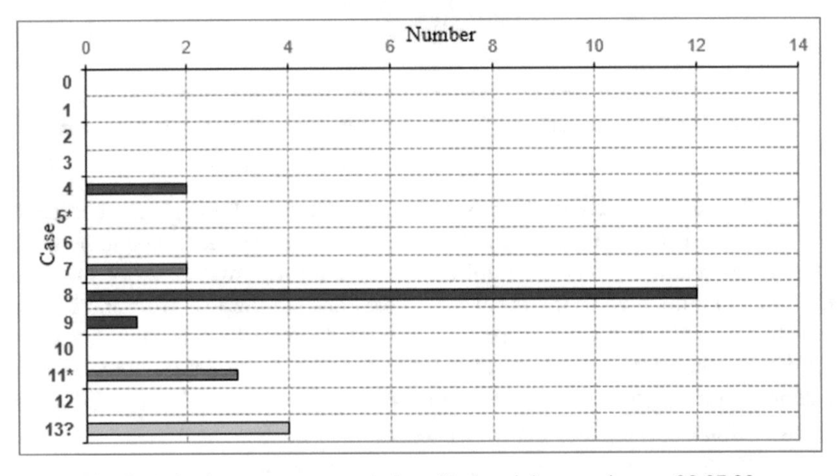

Fig. 8. Joint forecast cases statistics of industrial enterprise per 28.07.22.

Similarly, this situation is illustrated by the example of the forecast for the whole month (Fig. 9). Positive effect is not achieved every day. However in accordance with the calculation of target function [8] the effect for the month is positive and amounts to 39.3 thousand rub. At the same time, reducing the forecast error by only 0.1% adds about 50% more to the month effect. Total effect for July 2023 is 57.4 thousand rub.

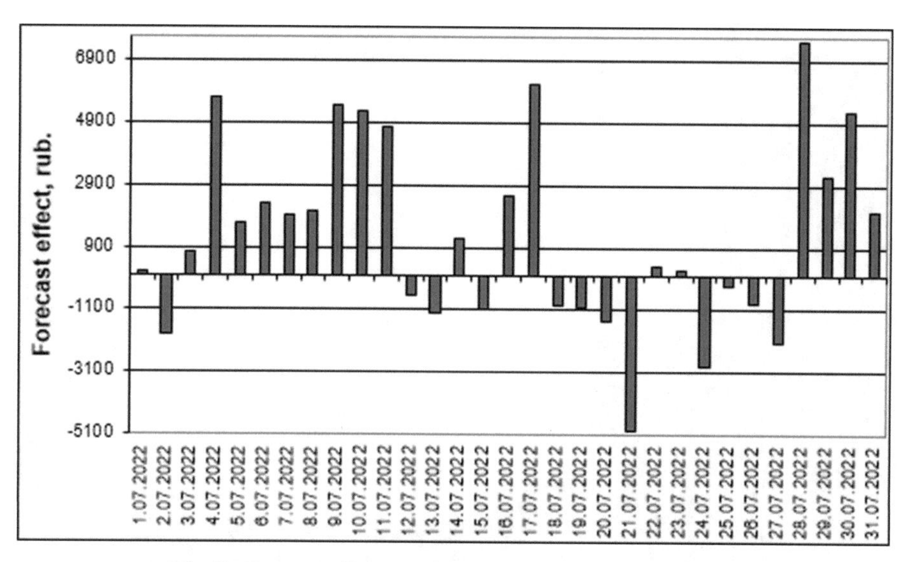

Fig. 9. Forecast effect of industrial enterprise per July 2022.

The distribution of different forecast cases [8] shows the predominance (Fig. 10) of cases that bring positive effect or reduce the forecast error (blue and green colors) to cases that bring loss (red color).

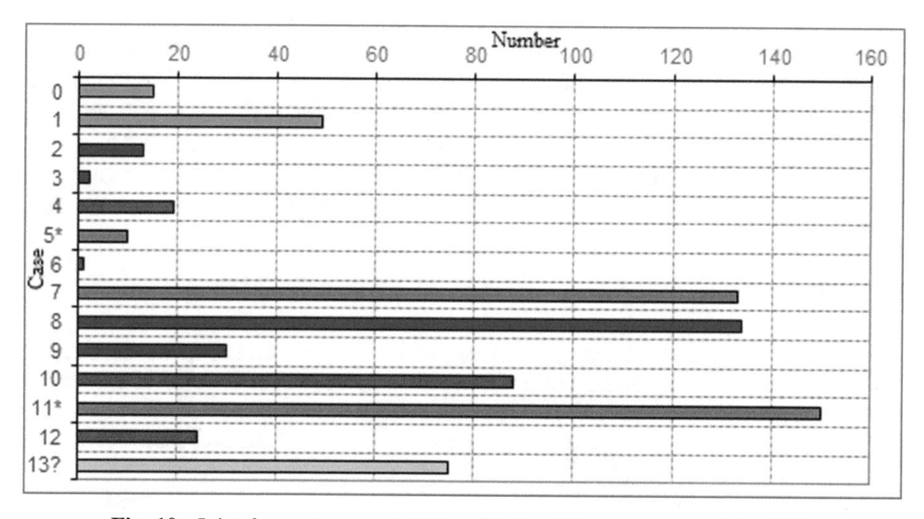

Fig. 10. Joint forecast cases statistics of industrial enterprise per July 2022.

The distribution of the target function (Formulas 2, 3) parts ratio in July 2022 is shown at Fig. 11.

The highest density of F_{1+}/F_{1-} ratio is in the region of 1.1–1.3. It explains the maximization of forecast effect with the ratio equals to 1.1 (Fig. 5).

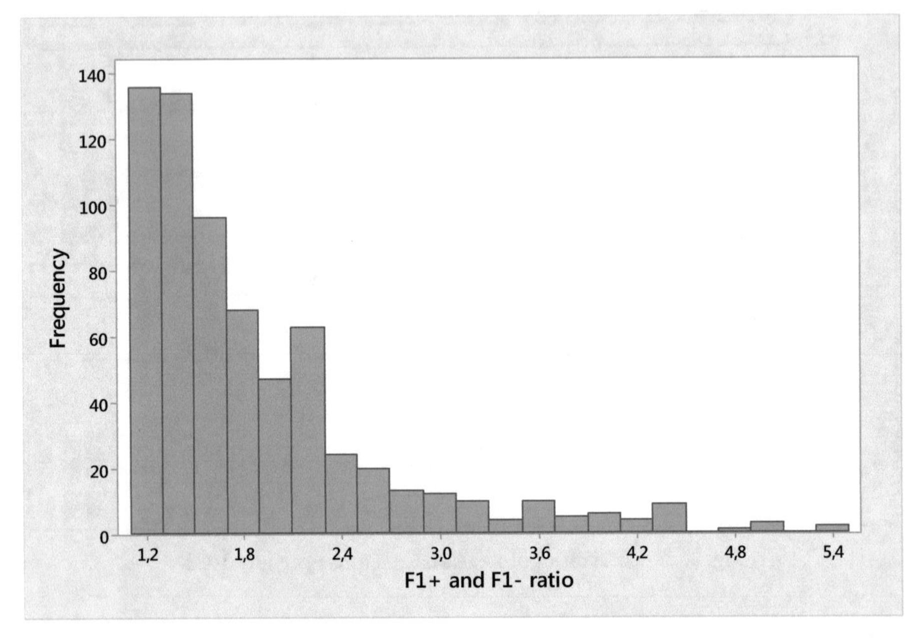

Fig. 11. Target function ratio of industrial enterprise per July 2022.

4 Conclusion

Described forecast model makes it possible to overcome the difficulties of existing forecast time series models (in particular, neural network and autoregressive) in forecasting uneven time series. Using the concept of "forecasting efficiency" instead of "forecasting error" more accurately reflects a comprehensive approach to joint forecasting of uneven and adjacent time series. Authors are continuously working to fine-tune and improve the forecast model. There are plans to refine the model for its application in stock markets.

References

1. Pirjan, A., Oprea, S.-V., Carutasu, G., Petrosanu, D.-M., Bara, A., Coculescu, C.: Devising hourly forecasting solutions regarding electricity consumption in the case of commercial center type consumers. Energies **10**(11), 17–27 (2017)
2. Weron, R.: Electricity price forecasting: a review of the state-of-the-art with a look into the future. Int. J. Forecast. **30**(4), 1030–1081 (2014)
3. Yang, Jing, F.: Power system short-term load forecasting. Thesis for Ph.D degree, p. 139. Elektrotechnik und Informationstechnik der Technischen Universitat, Darmstadt (2006)
4. Haykin, S.: Neural Networks. A Comprehensive Foundation, p. 842. Prentice Hall (1999)
5. Draper, N.R., Smith, H.: Applied Regression Analysis, p. 709. Wiley, New York (1981)
6. Russkov, O.V., Saradgishvili, S.E.: IT-method for uneven energy consumption planning. In: 2017 International Conference on Industrial Engineering, Applications and Manufacturing (ICIEAM), pp. 1–4 (2017)

7. Russkov, O.V., Saradgishvili, S.E.: The method of planning the energy consumption for electricity market. In: IOP Conference Series: Earth and Environmental Science, vol. 90, 012068 (2017)
8. Russkov, O.V., Saradgishvili, S.E.: A digital method for correcting planned electric energy consumption as a step to the Energynet market. In: SHS Web of Conference, vol. 44 (2018)
9. Harris, M.: Inside the Crystal Ball: How to Make and Use Forecasts, 397 p. Wiley, New York (2015)
10. Russkov, O., Saradgishvili, S.: The electricity market prices forecast as energy efficient procedure for an industrial monotown enterprise. Proc. Eng. **117**, 309–316 (2015)
11. Devaine, M., Gaillard, P., Goude, Y., Stoltz, G.: Forecasting electricity consumption by aggregating specialized experts. Mach. Learn. **90**(2), 231–260 (2012). https://doi.org/10.1007/s10994-012-5314-7
12. Ross, S.: A First Course in Probability, 8th edn., 530 p. Pearson Prentice Hall (2010)
13. von Neuman, J., Morgenstern, O.: Morgenstern Theory of Games and Economic Behavior, 674 p. Princeton University Press (1953)

Analysis of the Financial and Risk System of the Insurance Procurement Scoring Model

Evgenii Makarenko[1,2(✉)], Ekaterina Lukina[2], and Fazliddin Khujaev[3]

[1] St. Petersburg Polytechnic University of Peter the Great, 29, Polytechnicheskaya Str., St. Petersburg 195251, Russia
ss300@yandex.ru

[2] St. Petersburg State University of Aerospace Instrumentation, 67, Bolshaya Morskaya Str., St. Petersburg 190000, Russia

[3] Tashkent State University of Economics, 49, Islam Karimov Str., 100066 Tashkent, Uzbekistan

Abstract. The article analyses the main indicators of the scoring model for insurance services purchase as the authorized capital amount and the net profit of the insurer amount. We have analyzed the period of the Russian insurance market over the last ten years and found both the presence of correlation in some cases and the absence of a number of indicators, aiming at building this information dependency model. We have suggested a number of options for using the findings of our studies in order to build an effective digital procurement model that maximizes competitiveness and reduces the risks of corruption and lobbying. The risks of applying preferences in the scoring model for the procurement of insurance services for SMEs have also been considered. These procurements have been found particularly risky and not feasible. Conclusions have been drawn on ways of minimizing these risks in the digital procurement model.

Keywords: Insurance Market · Scoring Model · Authorized Capital · Profit · Insurance Company · Insurance Industry · Small and Medium-Sized Businesses · Losses

1 Introduction

At the moment, the situation of risk characterizes the decision to choose one insurer or another, for all customers. In Russia, there is no unified customer scoring system with interlinked components where customers could choose the necessary parameters and form both, their own system of risks to be insured and the system of criteria for selecting the insurance company. The formation of this digital system must be guided by guidelines based on formal logic and mathematics and the code must be completely transparent to all users.

The data from the Central Bank of the Russian Federation (the CBR) shows that the number of insurance market players in Russia continues to decline, as of the end of 2022 and premis continue to be concentrated among the leading companies, both in terms of collections and geographically in Moscow and St Petersburg. According to Insurance

© The Author(s), under exclusive license to Springer Nature Switzerland AG 2023
A. Gibadullin (Ed.): DITEM 2022, LNNS 683, pp. 52–63, 2023.
https://doi.org/10.1007/978-3-031-30926-7_6

Today and the Central Bank website, the number of insurance companies decreased from 520 to 136 (minus 73.8%) between 2012 and 2022 (includingly).

Therefore, the purpose of this paper is to develop the system of mathematical indicators for the scoring model that will enable the decision to choose a particular insurance company with a minimum risk of losing the license of a particular insurer.

2 Materials and Methods

The input data is the insurers' annual reports, the CBR's decisions on license revocation and insurers' decisions on transfer of insurance portfolios for the period from 2012 to 2022. The information from the Federal Tax Service's Unified Register of Small and Medium-Sized Enterprises for the period from 2016 to 2022 has also been used in the analyses. The statistical tools were used in the analysis to determine the closeness of association between the two traits, such as the Pearson association coefficient and the Yule-Kendall contingency coefficient.

3 Results and Discussion

The Russian legislation contains several options for the implementation of competitive procurement in terms of types of enterprises and forms of competitive procedures. At present, the main laws regulating procurement activities are Federal Law No. 44-FZ on the Contract System in the Sphere of Procurement of Goods, Works and Services for State and Municipal Needs (hereinafter referred to as 44-FZ), Federal Law No. 223-FZ on Procurement of Goods, Works and Services by Certain Types of Legal Entities (hereinafter referred to as 223-FZ) and Federal Law No. 135-FZ on Protection of Competition (hereinafter referred to as 135-FZ).

When evaluating the non-price criteria of providers, the customers build mathematical evaluation models containing criteria such as: authorized capital and net profit, and preferential treatment for small and medium-sized enterprise (hereinafter SME) companies may also be applied.

3.1 Assessment of the Insurer's Authorized Capital Amount

As it has been mentioned above, the size of the authorized capital is one of the main non-price criteria in choosing an insurance company. This is mainly due to the accessibility of this information, the ease with which it can be verified and the lack of the need for any complex mathematical calculations, as if it were necessary to calculate and verify, for example, a combined loss ratio.

Customers often use the amount of authorized capital as one of the non-price (financial) criteria for evaluating an insurance company in the risk management scoring model, along with the amount of net profit, level of benefits, etc. [2]. If put aside the lobbying of specific insurers and the corruption component of competitive procurement, the customer's logic is as follows: the larger the authorized capital of the insurer is, the lower the risks of revocation of the insurance license by the CBR are, the higher the probability of fulfilling the obligations is, both in case of the insurance events and ongoing support of the insurance contract (for example, amending the existing insurance policies) and the lower the risks of voluntarily surrendering the license in case the owners of the company are no longer interested in their business, for example as it happened to the Uralsib Insurance Group [1].

In terms of the risk management model, it is important to assess whether the amount of authorized capital has an impact on the probability of the insurer's license withdrawal [3]. Let us determine the strength of the relationship between the size of the authorized capital and the fact that insurance companies' licenses have been revoked in the Russian Federation between 2012 and 2022. The practical absence or weak relationship between the amount of authorized capital and the withdrawal of an insurance company's license is the hypothesis to be tested.

The data from the Russian Federal Tax Service was used to check the hypothesis of relationship between the size of the insurer's authorized capital and the probability of its exiting the market. The logic for obtaining information on the amount of authorized capital at the time of license revocation is shown in the Fig. 1. This is primarily due to the difficulty in retrospective estimating the amount of share capital at the time of license revocation due to imperfect information systems of the Central Bank of the Russian Federation and the Federal Tax Service.

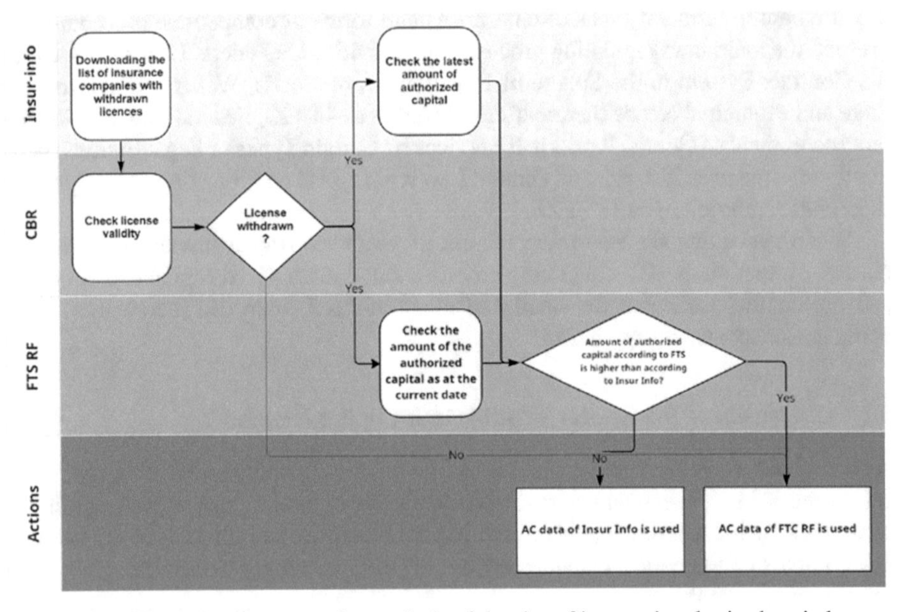

Fig. 1. The logic of retrospective analysis of the size of insurers' authorized capital.

Since the license withdrawal attribute is an alternative one and it involves only two variants of values, we use statistical tools to determine the strength of the relationship between the two attributes, such as the Pearson association coefficient, the Yule coefficient of colligation and the Yule-Kendall contingency coefficient [4, 5].

The universe of 282 insurance companies will be tentatively divided into four subgroups: two groups each according to the following attributes:

- The first group includes the companies with authorized capital of up to RUR 600 million (small companies) and the second group includes the companies with authorized capital exceeding RUR 600 million - medium and large companies. The amount of RUR 600 million was taken on the basis that this is the amount of share capital that will be needed from 2024 to provide the full range of services for both insurance and reinsurance.
- Bearing in mind the fact that a license has been withdrawn, we distinguish the two groups of companies: the first one includes insurance organizations that have lost their insurance license, and the second group includes organizations that continue their insurance activities in Russia.

Table 1 shows the distribution of Russian insurance companies by the size of authorized capital and presence/absence of an insurance license. The cells indicate the number of companies that match the combination of attribute values.

Table 1. Breakdown of Russian insurance companies by size of authorized capital and whether they have a valid insurance license.

The authorized capital amount, mln. RUR	Insurance license availability		TOTAL
	Available	Not available	
Less than 600	132 (a)	91 (b)	223
more than 600	31 (c)	28 (d)	59
TOTAL	163#	119	282

We calculate the association, contingent and colligation coefficients:

$$\text{K}_\text{a} = \frac{a \cdot d - b \cdot c}{a \cdot d + b \cdot c} = \frac{132 \cdot 28 - 91 \cdot 31}{132 \cdot 28 + 91 \cdot 31} = 0.13 \tag{1}$$

$$\text{K}_\text{K} = \frac{a \cdot d - b \cdot c}{\sqrt{(a+c) \cdot (b+d) \cdot (a+b) \cdot (c+d)}} = \frac{132 \cdot 28 - 91 \cdot 31}{\sqrt{163 \cdot 119 \cdot 223 \cdot 59}} = 0.06 \tag{2}$$

$$\text{K}_{\text{кл}} = \frac{\sqrt{a \cdot d} - \sqrt{b \cdot c}}{\sqrt{a \cdot d} + \sqrt{b \cdot c}} = \frac{\sqrt{132 \cdot 28} - \sqrt{91 \cdot 31}}{\sqrt{132 \cdot 28} + \sqrt{91 \cdot 31}} = 0.07 \tag{3}$$

The coefficient values indicate a very weak and virtually non-existent static relationship between the size of share capital and the fact that Russian insurance companies have had their licenses revoked.

We use Student's t-test to assess the significance of the correlation determined by the association coefficient:

$$t_r = \frac{k_a \cdot \sqrt{n-2}}{\sqrt{(1-k_a^2)}} = \frac{0,13 \cdot \sqrt{(282-2)}}{\sqrt{(1-0,13^2)}} = 2.591 \tag{4}$$

When the number of the degrees of freedom f = 280 and the significance level equal to 0.05, the t-test value (t_{test}) is 1,968. The calculated value of t_r (2.591) is greater than t_{crit} (1.968), so the obtained coefficient is statistically significant and the hypothesis of a very weak relationship between the attributes is confirmed.

To test the robustness of the tested statistical hypothesis, we shift the boundary of the distribution of insurance companies by authorized capital to RUB 500 million, since the universe of insurance companies includes a larger number of small companies [6]. In this case the boundary would be more in line with the actual median rather than the legal status (see Table 2).

Table 2. Breakdown of Russian insurance companies by size of authorized capital and bankruptcy.

The authorized capital amount, mln. RUR	Insurance license availability		TOTAL
	Available	Not available	
Less than 500	129 (a)	76 (b)	205
more than 500	34 (c)	43 (d)	77
TOTAL	163	119	282

We calculate the association, contingent and colligation coefficients:

$$K_a = \frac{a \cdot d - b \cdot c}{a \cdot d + b \cdot c} = \frac{129 \cdot 43 - 76 \cdot 34}{129 \cdot 43 + 76 \cdot 34} = 0.36 \quad (5)$$

$$K_k = \frac{a \cdot d - b \cdot c}{\sqrt{(a+c) \cdot (b+d) \cdot (a+b) \cdot (c+d)}} = \frac{129 \cdot 43 - 76 \cdot 34}{\sqrt{163 \cdot 119 \cdot 205 \cdot 77}} = 0.17 \quad (6)$$

$$K_{кл} = \frac{\sqrt{a \cdot d} - \sqrt{b \cdot c}}{\sqrt{a \cdot d} + \sqrt{b \cdot c}} = \frac{\sqrt{129 \cdot 43} - \sqrt{76 \cdot 34}}{\sqrt{129 \cdot 43} + \sqrt{76 \cdot 34}} = 0.19 \quad (7)$$

The coefficient values indicate that there is little correlation between the size of authorized capital and the fact that an insurance company's license has been withdrewn in Russia, the same as a low size of authorized capital does not increase the risk of insurance companies losing their license, nor does a high size of authorized capital guarantee the insurer's sustainable solvency.

We use Student's t-test to assess the significance of the correlation determined by the association coefficient:

$$t_r = \frac{k_a \cdot \sqrt{n-2}}{\sqrt{(1-k_a^2)}} = \frac{0{,}36 \cdot \sqrt{(282-2)}}{\sqrt{(1-0.36^2)}} = 6.452 \quad (8)$$

When the number of the degrees of freedom f = 280 and the significance level equal to 0,05, the t-test value (t_{test}) is 1.968. The calculated value of t_r (6.452) is greater than t_{crit} (1.968), so the obtained coefficient is statistically significant and the hypothesis of weak relationship between the attributes is confirmed.

Federal Law No. 327-FZ On Amendments to Certain Legislative Acts of the Russian Federation establishes that insurance companies (other than insurers of obligatory medical insurance (CMI) must possess authorized capital of at least RUR 300 million by 1 January, 2024, not less than RUR 450 million for life insurance, and not less than RUR 600 million for reinsurance.

As a result, 27 of the remaining 132 participants in the Russian market (which is 20.5% of the Russian market) with a total capital shortfall of RUR 1.6 billion this or that extent do not meet the necessary requirements of the Central Bank.

Currently, the minimum authorized capital requirement for insurers under the 2024 law should be one of the key requirements imposed by customers when preparing procurement documents for insurance services, both under 44-FZ and 223-FZ, as well as in any other competitive procurement. It should be pointed out that relatively few insurance companies fall under this restriction. In addition, it does not violate the principles of 135-FZ, ensuring that unfair competition is avoided and that conditions are created for the effective functioning of the insurance services market [19] and should therefore be the key one in the scoring model for assessing insurers.

3.2 Estimating the Insurer's Profit Margin

Profit is one of the main indicators of an insurance company's financial performance. It is the basis of economic development, as profit creates the financial basis for internal self-financing, expansion of the insurance business, technical re-equipment and innovation, and the solution of social and material issues of the insurer's staff. Correspondingly, the higher the profit is, the more efficient is the insurer's activity, the scarcer is the probability of the insurer's bankruptcy, a lot of insurers include the net profit amount into the scoring system of non-pricing rates of the providers evaluation [7, 18].

We determine the strength of the relationship between the existence of economic and financial losses and the fact losing the licence by the insurer in the Russian Federation between 2018 and 2022. The hypothesis to be checked is that there is a close statistical relationship between economic and financial losses and the revocation of an insurer's insurance license.

Since the presence of losses attribute and license withdrawal attribute are alternative ones and it involves only two variants of values, we use statistical tools to determine the strength of the relationship between the two attributes, such as the Pearson association coefficient and the Yule-Kendall contingency coefficient.

The universe of 150 insurance companies will be tentatively divided into four subgroups: two groups each according to the following attributes:

1. Bearing in mind the presence of losses, we distinguish two groups of companies: the first group includes organizations that have losses and the second one includes organizations that have not had losses.
2. Bearing in mind the fact that a license has been lost, we distinguish the two groups of companies: the first one includes the organizations that have lost their license, and the second group includes organizations that continue their insurance activities.

Table 3 shows the distribution of Russian insurance companies by economic loss and bankruptcy. The cells indicate the number of companies that match the combination of attribute values.

Table 3. Breakdown of Russian insurance companies by presence of losses and bankruptcy.

Existence of a loss for the year	Insurance license availability		TOTAL
	Available	Not available	
Yes	21 (a)	13 (b)	34
No	26 (c)	124 (d)	150
TOTAL	47	137	184

We calculate the association, contingent and colligation coefficients:

$$K_a = \frac{a \cdot d - b \cdot c}{a \cdot d + b \cdot c} = \frac{21 \cdot 124 - 13 \cdot 26}{21 \cdot 124 + 13 \cdot 26} = 0.77 \tag{9}$$

$$K_K = \frac{a \cdot d - b \cdot c}{\sqrt{(a+c) \cdot (b+d) \cdot (a+b) \cdot (c+d)}} = \frac{21 \cdot 124 - 13 \cdot 26}{\sqrt{47 \cdot 133 \cdot 34 \cdot 150}} = 0.4 \tag{10}$$

The coefficient values indicate that there is a significant static relationship between the size of authorized capital and the fact of bankruptcy of Russian insurance companies.

We use Student's t-test to assess the significance of the correlation determined by the association coefficient:

$$t_r = \frac{k_a \cdot \sqrt{n-2}}{\sqrt{(1-k_a^2)}} = \frac{0.77 \cdot \sqrt{(184-2)}}{\sqrt{(1-0.77^2)}} = 16.281 \tag{11}$$

When the number of the degrees of freedom f = 182 and the significance level equal to 0.05, the t-test value (t_{test}) is 1.973. The calculated value of t_r (16.281) is greater than t_{crit} (1.973), so the obtained coefficient is statistically significant and the hypothesis of strong correlation relationship between the attributes is confirmed.

To check the hypothesis about the loss size influencing the subsequent decision by the CBR to withdraw an insurance company's license, we divide the insurance companies that have suffered a loss into two groups based on the loss size (see Table 4).

Table 4. Breakdown of Russian insurance companies by size of loss and license withdrawal.

Amount of loss for the year, RUB mln	Insurance license availability		TOTAL
	Available	Not available	
Less than 250	19 (a)	10 (b)	29
more than 250	2 (c)	3 (d)	5
TOTAL	21	13	34

We calculate the association, contingent and colligation coefficients:

$$K_a = \frac{a \cdot d - b \cdot c}{a \cdot d + b \cdot c} = \frac{19 \cdot 3 - 10 \cdot 2}{19 \cdot 3 + 10 \cdot 2} = 0.48 \tag{12}$$

The coefficient values indicate a weak relation which is tending towards average statistical relationship between the size of the loss and the fact of insurer's license withdrawal by the Central Bank of Russia.

We use Student's t-test to assess the significance of the correlation determined by the association coefficient:

$$t_r = \frac{k_a \cdot \sqrt{n-2}}{\sqrt{(1-k_a^2)}} = \frac{0.48 \cdot \sqrt{(34-2)}}{\sqrt{(1-0.48^2)}} = 3.095 \tag{13}$$

When the number of the degrees of freedom f = 32 and the significance level equal to 0.05, the t-test value (t_{test}) is 2,036. The calculated value of t_r (3.095) is greater than t_{crit} (2.036), so the obtained coefficient is statistically significant and the hypothesis of weak correlation relationship between the attributes is confirmed.

3.3 Assessment of an Insurer's Affiliation to a Small or Medium-Sized Business

In Russia, since 1 August, 2016 a unified register of small and medium-sized enterprises (hereinafter referred to as SMEs) has been maintained in accordance with Federal Law No 408-FZ "On Amendments to Certain Legislative Acts of the Russian Federation".

The main criteria for classifying an insurance organization as a small or medium-sized enterprise are laid down in the Federal Law on the Development of Small and Medium-Sized Entrepreneurship in the Russian Federation. To this end, the following is taken into account:

- The average number of employees in the previous calendar year;
- The amount of business revenue for the previous calendar year;
- The authorized capital structure.

Being on the SME register allows businesses to obtain a number of preferences and support options.

There are 22 SME insurance companies registered in Russia as of December 2022. The proportion of SME insurers among the insurance companies that have had their licenses withdrawn is disproportionately higher and varies annually from 26 to 40%, at the end of 2022 it was 29%, i.e. almost every third company that had its license withdrawn was an SME (Fig. 2) [15]. It means that the probability of a SME insurer losing its license is significantly higher than that of an ordinary insurer [12]. At that, it may be noted that the situation is such that only one SME company in Russia is licensed for such types of insurance as life insurance, compulsory motor third party liability insurance (CMTPL) and compulsory third party liability insurance for damage caused by an accident at a hazardous facility (MTPL) [16, 17].

Customers making procurements under 44-FZ are required to make procurements from small businesses and socially oriented non-profit organizations in an amount of at least 25% of the total annual volume of procurements. However, in the procurement notice and documentation, the customer points out that the procurement has been placed for a certain category of providers, which actually means that the representatives of other organizations cannot participate in such a procedure (the electronic platform simply will not allow this to happen) [20]. Failure to comply with this SME procurement volume rule is considered an administrative offence for which a fine is imposed. This leads to trying to make a large variety of purchases from SMEs [8], which in turn carries risks that are visualized in the butterfly diagram in Fig. 2.

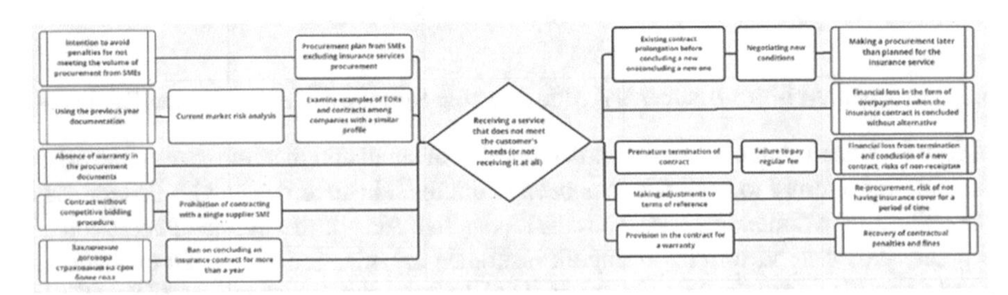

Fig. 2. Risk butterfly diagram for the procurement of insurance services from SMEs.

All risks can be grouped into the following categories:

1. Financial risks in the performance of the contract

1.1. A contract performance security is a monetary obligation provided by the successful bidder when the contract is awarded to cover possible damage to the customer if the supplier fails to perform or perform properly. It may be between 0.5% and 30% of the initial maximum contract price (hereinafter referred to as the MIPC) (this may be either money which is transferred to the customer's account or an independent guarantee). At that, if the insurer is an SME representative, the participant is entitled to provide evidence of the necessary experience, i.e. without financial guarantees on their part. The lack of financial warranties leads to the impossibility of obtaining any compensation in case of early termination of the contract or non-fulfilment of the obligations by the contractor [13].

1.2. Warranty obligations are mandatory warranty service for the service provided. The provider shall provide security for such obligations either in money, which shall be frozen for the duration of the warranty service, or in an independent guarantee not exceeding 10% of the MSC. In the case of procurement from SMEs, the provider may be exempted from providing such warranties, if such a participant in the procurement provides the information contained in the register of contracts and confirming that such participant executed three contracts without penalties (fines, penalties) being applied to such participant within three years before the date of submission of the procurement. Similarly to the above paragraph the lack of warranty reduces the likelihood of financial compensation in case of breaches of contract.

1.3. Increasing insurance costs. Due to the small number of SME insurance companies in Russia and the fact that only a few of them are actively involved in public procurement, the reality is that in some cases the procurement is made at an arm's length price, without any price reduction, as there is only one company involved in the procurement. This eventually leads to cost overruns on the part of the customer [9].

2. The risk of contract non-performance
The distinctive feature of the insurance services is that their probability of use is not 100%, but in many cases even less than 1% (especially property and liability insurance), and the validity of an insurance policy is usually at least one year. However, most purchasers treat their procurement with less responsibility than, for example, the procurement of site security services. Procurement of insurance services from SMEs significantly increases the risk of the provider losing its insurance license and of the customer being left without the insurance coverage at all, which may at least lead to a new unplanned procurement of insurance services, i.e. to the overspending of the insurance budget at the expense of other items or losses that the customer will have to cover himself.

3. Risk of delay in the procurement process
Among the insurance services available for the procurement there are specific types of insurance that small insurance companies do not normally handle, such as shipowner's liability insurance. However, customers try to place orders for these types of insurance among SMEs, which results in the procurement being declared unsuccessful and the customer having to re-advertise the procurement to a larger number of bidders [10].

From the point of view of the 223-FZ, there is no particular difference from the 44-FZ in terms of procurement from SMEs, the only difference being that the 25% procurement volume must not only be from small businesses but also from medium-sized businesses, which slightly increases competition among contractors, but does not change the overall risk picture much [11].

4 Conclusion

1. The hypothesis that there is no significant relationship between the amount of authorized capital and license withdrawal from insurance companies has been confirmed by correlation analysis.
2. The results of the correlation analysis are significant, as confirmed by calculating Student's t-test.
3. The hypothesis has proved to be robust to changes in the frontier distribution of insurance companies in terms of authorized capital.

It can be stated that a small authorized capital does not increase the probability of an insurer having its license withdrawn, just as a high authorized capital does not guarantee their sustainable solvency and guarantee their continued presence in the insurance market. Further tightening of the CBR's requirements for the financial stability of insurers will lead to further increase of charter capital requirements. In this regard, customers need to take into account and monitor prospective requirements of the CBR [12] for authorized capital in scoring models and avoid entering into both state contracts and other risk insurance contracts with the insurers that do not meet the future requirements of the CBR, while the setting authorized capital as one of the non-price criteria for assessing the potential customer's financial condition or qualifications is more a lobbying of large insurers than an economically and scientifically stated way of protecting the customer from contracting an untrustworthy provider.

4. The link between the losses and the probability of an insurer losing its license has been confirmed, but in the procurement scoring model this criterion should be used by actual loss/gain for the period, rather than by reference to profit or loss.
5. Applying preferential treatment to SMEs increases the financial risk for the customer.

Undoubtedly refusal (or ban) to procure insurance services from SMEs is the most appropriate solution. We have be taken into account the fact that the further reduction of SME insurance companies number will increase the risk of appeals to the FAS against such procurement by other insurance customers, when forming the mathematical evaluation criteria for performers.

If such a procurement is unavoidable for one reason or another, at least the procurement should take place a long time before the insurance contract is due to start. This will allow for time to re-advertise the procurement in case there are no SMEs interested. It should also be possible to include guarantee clauses in the tender documentation, which would reduce the attractiveness of the procurement in the eyes of unreliable suppliers on the one hand and somewhat safeguard the customer on the other hand [14].

On the one hand, the use of these findings in the scoring model for assessing potential providers of insurance services significantly expands the range of potential bidders by increasing competition among providers and, on the other hand, it reduces the risk of unreliable providers among SMEs being awarded contracts.

References

1. Barykin, S.Y., et al.: The sharing economy and digital logistics in retail chains: opportunities and threats. Acad. Strat. Manag. J. (2021)
2. Barykin, S.Y., Kapustina, I.V., Valebnikova, O.A., Valebnikova, N.V., Kalinina, O.V., Sergeev, S.M., Camastral, M., Putikhin, Y., Volkova, L.: Digital technologies for personnel management: implications for open innovations. Academy of Strategic Management Journal (2021)
3. Belozyorov, S.A., Sokolovska, O.: Economic sanctions against Russia: assessing the policies to overcome their impact. Econ. Reg. **16**(4) (2020)
4. Belozerov, S., Sokolovskaya, E.: The game-theoretic approach to modeling the conflict of interests: the economic sanctions. Terra Econ. **20**(1), 65–80 (2022)
5. Bril, A.R., Kalinina, O.V., Ilin, I.V.: Financial and economic aspects of IT project management. In: Proceedings of the 30th International Business Information Management Association Conference, IBIMA 2017 - Vision 2020: Sustainable Economic development, Innovation Management, and Global Growth (2017)
6. Chernogorskiy, S.A., Kozlov, A.V., Teslya, A.B.: Game-theoretic modeling of decision-making on state support for the infrastructure development in the Russian Far North. Int. J. Syst. Assur. Eng. Manag. **11**(1), 10–18 (2019). https://doi.org/10.1007/s13198-019-00798-6

7. Gromova, E.A., Pupentsova, S.V.: Simulation modelling as a method of risk analysis in real estate valuation. IOP Conf. Ser. Mater. Sci. Eng. **898**(1) (2020)
8. Ilin, I., Klimin, A., Shaban, A.: Features of big data approach and new opportunities of BI-systems in marketing activities. In: E3S Web of Conferences, p. 110 (2019)
9. Kerimov, M., Belinskaia, I., Evdokimov, K., Samorukov, V., Klochkov, Y.: Technological machines operation by identification method. Int. J. Math. Eng. Manag. Sci. (2022)
10. Klochkov, Y.S., Tveryakov, A.M.: Approaches to the improvement of quality management methods. Int. J. Syst. Assur. Eng. Manag. (2020)
11. Klochkova, E., Evdokimov, K., Klochkov, Y., Samorukov, V.: Methodology for reducing risk of underperformance of personnel functions. In: Engineering for Rural Development. Proceedings (2018)
12. Kopteva, L., Budagov, A., Trushevskaya, A.: Impact of corruption on the economic and environmental security of the state. In: E3S Web of Conferences (2021)
13. Krichevskiy, M.L., Dmitrieva, S.V., Martynova, Y.A.: Neural network assessment of personnel competences. Labor Economics (2018)
14. Makarenko, E., Lukina, E., Pletneva, N.: Mathematical modeling of monopolization processes in the market of compulsory carrier liability insurance. Transp. Res. Procedia **12** (2022)
15. Parfenova, V.E., Bulgakova, G.G., Amagaeva, Y.G., Evdokimov, K.V.: Fuzzy modeling for task of management of the agricaltural-industrial complex. In: IOP Conference Series: Materials Science and Engineering (2019)
16. Pupentsova, S., Livintsova, M.: The enterprises risk management in the context of digital transformation. In: Manakov, A., Edigarian, A. (eds.) TransSiberia 2021. LNNS, vol. 403, pp. 1159–1167. Springer, Cham (2022). https://doi.org/10.1007/978-3-030-96383-5_129
17. Pupentsova, S., Livintsova, M., Shabrova, O.: A model for determining the market rental rate for properties classified in the segment with a limited number of offers. IBIMA Bus. Rev. (2021)
18. Schepinin, V., Bataev, A.: Digitalization of financial sphere: challenger banks efficiency estimation. IOP Conf. Ser. Mater. Sci. Eng. **497**(1) (2019)
19. Svirina, A., Appalonova, N., Garanin, D., Lukashevich, N., Koshkin, I.: Fintech developmental trends: the role and influence of sustainable digital logistics. In: E3S Web of Conferences, vol. 258 (2021)
20. Wentao, W., Makarenko, E.: Adaptation of the experience of digitalization of the Chinese insurance industry in the favor of the development of technologies of the Russian insurance market. Digital and information technologies in economics and management. In: Proceedings of the International Scientific and Practical Conference "Digital and Information Technologies in Economics and Management" (DITEM2021) (2022)

Innovative Model, Social Financial Technologies and Sovereign Issue as a Tool for Accelerated Development of the Elkon Mining and Metallurgical Combine

Evgeniy V. Kostyrin(✉) and Evgeniy V. Sokolov

Bauman Moscow State Technical University, Building 1, 5, 2-Nd Baumanskaya Str., Moscow, Russia

kostyrinev@bmstu.ru

Abstract. The article has developed an economic and mathematical model combining sovereign emission, a comprehensive system of social financing of enterprises and optimizing the volume of interest-free government loans, the amount of credit funds returned to the state (the body of the loan), wages of the labor collective, consistent with revenue growth, deductions for the development of the enterprise, taxation and social contributions. Using sovereign emission and social financial technologies, the development process from 2023 to 2033 (10 years) is modeled on the joint-stock company "Elkon Mining and Metallurgical Combine". As a result of interest-free government lending and social financial technologies for the enterprise in question, revenue increased 2.05 times over 10 years, wages 8.6 times, development contributions 4.76 times. As a result, in the third year of the simulation, the volume of tax revenues will exceed the loan body by 173 million rubles, for the 5th year by 382 million rubles, for the 10th year by 1,463 million rubles (3 times). In other words, the state, providing interest-free loans to the enterprise and completely writing off the loan debt (the body of the loan), receives 3 times more funds to the budget than it invests in the enterprise.

Keywords: Sovereign Issue · Interest-Free Government Loans · Wages · Economic and Mathematical Model · Social Financial Technologies · Import Substitution

1 Introduction

In this article, using the example of JSC "Elkon Mining and Metallurgical Combine", the use of a model of social financial technologies for the development of enterprises is considered [1]. This article can be found on the sokolov.expert website in the "Science" section. Based on the data of the balance sheet, cash flow statement, financial results report and the number of employees of the selected organization, starting from 2023, an enlarged forecast of development has been developed using an economic and mathematical model, algorithm and software.

© The Author(s), under exclusive license to Springer Nature Switzerland AG 2023
A. Gibadullin (Ed.): DITEM 2022, LNNS 683, pp. 64–76, 2023.
https://doi.org/10.1007/978-3-031-30926-7_7

Social technologies of enterprise financing are understood as labor relations harmoniously combining the financial interests of employees, owners and the state.

Social is related to people's lives and their relationships in society. Everything necessary (goods, works, services) for the life of people in society in any state is created in the process of labor. This can be work at enterprises of any form of ownership and scale (large, medium, small), the work of self-employed citizens, work within households (raising children, cooking, cleaning, etc.). In other words, the main social (public relations) necessary for a decent life of people are concentrated in the field of labor relations.

The earliest form of social protection of working citizens is collective insurance. *It appeared in Europe in the second half of the XIX century together with trade unions defending the rights of employees before employers and the state.* A typical example in this regard is England, where social security was implemented as a collective self-help and self-insurance by creating mutual aid funds, hospital funds, unemployment funds.

State social insurance, introduced in Germany in 1883 as part of the social reforms carried out by Bismarck, was the first legislative solution to the issue of social protection of the population in history. Laws were issued on sickness insurance, then on disability and old age insurance. These types of insurance were to be carried out by hospital funds, enterprise funds and free mutual assistance funds.

The beginning of the Russian social insurance system was laid in 1861, when the law "On the mandatory establishment of auxiliary partnerships at state-owned mining plants" was adopted. Insurance funds were formed from workers' contributions (2–3% of salary) and contributions of the plant management in an amount equal to the annual amount of workers' contributions. The collected funds were used to pay sickness benefits, pensions to the disabled, widows and orphans. In 1912, the III State Duma adopted a package of laws that laid the foundation for Russian social insurance: "On the approval of the presence of workers' insurance", "On the approval of the Council for Workers' Insurance", "On the provision of workers in case of illness", "On the insurance of workers from accidents at work".

The Great October Socialist Revolution and the very fact of the existence and dynamic development of the USSR had a huge impact on the accelerated development of State Social Insurance. When employers and the state clearly saw that if they do not allocate funds for social insurance of employees, they can be forcibly deprived of all their property and power.

Currently, all enterprises in Russia are required to make the following social contributions from the wages of employees: to the Pension Fund of Russia (PFR) – 22%; to the Compulsory Medical Insurance Fund (CMIF) – 5.1%; to the Social Insurance Fund (SIF) – 2.9% (in case of temporary disability and maternity) and from 2% - 8% (in case of injury).

All three social funds affect the interests of various groups of the Russian population. CMIF is used by all citizens of Russia (144 million people); PFR are citizens of working age (82 million people transfer funds to the PFR for them) and pensioners (36 million people). Total – 118 million people. FSS – citizens of working age – 82 million people.

Social technologies of enterprise financing are based on the development of economic and mathematical models, algorithms and tools (Excel, Mathcad and others) and include two main tasks:

Optimal distribution and use of social contributions of enterprises (PFR, CMIF, SIF).
The optimal ratio in enterprises of the financial interests of working citizens (labor collective), employers (owners) and the state.

The authors see the solution of the first task in the transition of healthcare and pension provision of Russian citizens to personalized medical savings accounts and personalized pension accounts and outlined the projects of such a transition in articles [2–4] and the draft federal law on amendments to the federal law "On Compulsory Medical Insurance in the Russian Federation", which was sent to the State Duma and with which you can find it on the sokolov.expert website, section "Science".

Social contributions in case of temporary disability and maternity and in case of injury are currently transferred by enterprises directly to those working at these enterprises, and the SIF receives the remnants of unspent funds, which, in case of a shortage of funds at enterprises, are paid to employees. A significant difference between the SIF and the PFR and the CMIF is that this fund is spent on citizens of working age and does not require savings for serving citizens of retirement age as the PFR and children and pensioners as the CMIF. Therefore, in the system of social financing of enterprises and the Russian economy proposed in this article, the functioning of the SIF does not change, that is, this fund operates according to the same legislative and regulatory acts as at present.

The second task of optimizing social (public) financial interests within the labor collectives of enterprises, where, as shown above, all material goods (goods, works, services) are created and from where budgets of all levels are filled.

Sovereign issue refers to the issue of money in circulation directed in the form of loans to enterprises producing import-substituting products or products for which demand is guaranteed.

On March 16, 2021, the FAIR RUSSIA – FOR TRUTH Party made concrete proposals aimed at establishing the self-sufficient nature of the Russian economy in the face of increased sanctions pressure on Russia and possible escalation of crisis phenomena in the economy. The package of measures developed by the Socialists is spelled out in the framework of the new economic program "COURSE", which stands for the Concept of Strengthening Russian Sovereignty.

One of the key proposals of this course is *that the Russian Government and the Central Bank organize the financing of enterprises not on a residual basis* with a lifting of the loan rate, but so that the rate for them is reduced from almost zero to 0.25%, *as it is done in Europe and the USA. In addition, we propose to make it possible for Russian enterprises to receive direct financing from the Ministry of Finance of the Russian Federation or the Central Bank, without involving commercial banks that look after their interests and want to profit from this money.* At the same time, it is possible to finance enterprises not just at a zero rate, but even write off part of their investment costs when expanding production, purchasing equipment, etc. This is made, for example, in China. There, when expanding production, increasing productivity, purchasing new equipment, money is given out at a zero rate, and then when the enterprise fulfills its obligations, part of these amounts is written off. With sovereign (project) financing in China, hundreds of millions of dollars are allocated for one enterprise. If the planned indicators are fulfilled within three years, the state writes off 30% of the subsidy, and in a year – another 20%. And we are talking about the body of the loan. In fact, the company increases productivity, receives new equipment for half its cost. This allows China to increase production capacity, accelerate economic development, and ultimately fill the country's budget. Moreover, this applies not only to enterprises affected by sanctions, but to anyone who declares their readiness to develop a particular production on our territory.

The approach proposed in the article, consisting in the sovereign issue of import-substituting enterprises, is harmoniously combined with the social financial technologies developed at the Department of Finance of the Bauman Moscow State Technical University for the development of enterprises and the economy of Russia [1]. Since in these approaches, the wages of working citizens are growing dynamically.

In the production of goods, the performance of works and the provision of services, the labor of working citizens (live labor) is spent at each workplace and materialized or past labor is used, which has passed into the cost of buildings, equipment, raw materials, materials, electricity and other elements of materialized labor used at enterprises. In other words, live (working citizens) and materialized labor is used at every workplace of any enterprise. *Accordingly, for the development of enterprises and the economy of the state, financial resources should first of all be directed to improving the professionalism and motivation of workers for high-performance work (wage growth) and to improving the equipment of workplaces (equipment, adaptations, working conditions, etc.) along all production chains.*

What is the current situation in Russia with the motivation of workers and highly effective work? Since 2014, the real incomes of the population have been falling annually. This is due to the fact that with the current taxation system and the collection of social payments to the owners of enterprises, it is *absolutely not profitable to increase wages, since this significantly increases the value added tax (at least 20% of wages) and social deductions (30% of wages).* In other words, by increasing the salary, the owner is forced to transfer 50% of it to the state. Therefore, it is much more profitable for the owner to pay the minimum wage (and the rest in the "envelope") or not to show it at all for some of the workers. Approximately 39 million of the working-age population in 2020 do not make any social contributions.

The main share of social contributions is made up of contributions to the pension fund (22% of wages). The situation with transfers of funds to the Pension Fund does not motivate working citizens to high-performance work, since the funded part of the pension has been frozen since 2014. The amount of the insurance pension that a working citizen will receive is still not clearly defined (the notorious individual pension coefficient). In the event of the death of a pensioner (say, a year after retirement), his insurance pension, *which he transferred to someone for about 40 years throughout his career*, is not inherited and, therefore, cannot improve the welfare of his family. It is also important to note that the Pension Fund was established in 1990 (*more than 30 years have passed*), but the situation with pensions *is not improving. The average insurance pension in Russia currently amounts to 16,905 rubles* [5], *which is only slightly higher than the national subsistence minimum of 13,919 rubles* [6].

Since contributions to the Pension Fund practically do not reach working citizens, approximately 39 million working citizens either do not transfer anything to social funds at all, or work in "gray" (officially receive the minimum wage, and the rest is in envelopes), which allows them to put "gray" money in the bank, buy real estate and thereby ensure a decent old age. At the same time, they receive the same pension and medical care as all Russian citizens. *It should be emphasized that "gray" wages are also beneficial for employers*, since the amount of social contributions and, consequently, the cost of production is significantly reduced. From the above, it should be concluded that *the existing system of taxation and financing of social security is not effective and forces working citizens and employers to go into the "shadow"*.

With such a growing trend in the number of working citizens who do not transfer anything or receive "gray" wages, the financing of pension provision and medical care *will only decrease.*

The Department of Finance of Bauman Moscow State Technical University has developed technologies for financing pension provision and medical care using the experience of Singapore, the USA, China and South Africa, according to which enterprises transfer social payments not to the Pension Fund and the compulsory Medical Insurance Fund, but to personalized pension and medical accounts placed in banks citizens working at these enterprises.

2 Materials and Methods

The economic and mathematical model [7, 8] combines a sovereign issue, a comprehensive system of social financing of enterprises and optimizes the volume of interest-free state loans, the amount of credit funds returned to the state (the body of the loan), the wages of the labor collective, consistent with revenue growth, deductions for the development of the enterprise (relevant for the employer and the entire labor collective), taxation and social contributions (important for the state), has the form:

Target function

$$W = R \cdot \theta_b + \xi \cdot (FR - FR_b) \rightarrow max, \quad (1)$$

Restrictions

$$C_{dev} = FR_b + (1 - \xi) \cdot (FR - FR_b) \cdot \left(1 - T_{profit}\right), \quad (2)$$

$$\theta = (R \cdot \theta_b + \xi \cdot FR)/R_b, \quad (3)$$

$$\Delta C = V \cdot \left(C_{var} + \frac{C_{fix}}{\sum_{i=1}^{n} V_i}\right) - V_b \cdot \left(C_{var} + \frac{C_{fix}}{\sum_{i=1}^{n} V_i}\right), \quad (4)$$

$$\Delta T = \Delta W \cdot T_{inc} + \Delta FR \cdot T_{profit} + (\Delta R - \Delta C_{var}) \cdot T_{VAT}, \quad (5)$$

$$FR = R - V \cdot \left(C_{var} + \frac{C_{fix}}{\sum_{i=1}^{n} V_i}\right), \quad (6)$$

$$\omega_{fix} = \frac{\frac{C_{fix}}{\sum_{i=1}^{n} V_i}}{C_{var} + \frac{C_{fix}}{\sum_{i=1}^{n} V_i}}, \tag{7}$$

$$\omega_{var} = \frac{C_{var}}{C_{var} + \frac{C_{fix}}{\sum_{i=1}^{n} V_i}}, \tag{8}$$

$$Credit = C_{sum} - C_{sumb}. \tag{9}$$

In the economic and mathematical model (1)–(9), the following designations are used: W - the amount of wages of working citizens, rubles; W_b – the amount of wages of working citizens in the basic version of the simulation, rubles; R – the income of enterprises from the sale of goods, products, works, services, rubles; θ – the percentage of income directed to increase the wages of working citizens; ξ – the coefficient of redistribution of the increase in the financial result between working citizens and owners of enterprises; ΔC – cost reduction due to the increase in sales of goods, products, works, services, rubles; C_{dev} – the amount of deductions allocated for the development of enterprises, rubles; R_b – the income of enterprises from the sale of goods, products, works, services in the basic version of the simulation, rubles; θ_b – the percentage of income directed to increase the wages of working citizens in the basic version of the simulation; V – the volume of sales of goods, products, works, services by enterprises, units; V_b – the volume of sales of goods, products, works, services by enterprises in the basic version of modeling, units; C_{var} – conditionally variable costs of enterprises in the sale of goods, products, works, services, rubles; C_{fix} – conditionally fixed costs of enterprises in the sale of goods, products, works, services, rubles; C_{sum} – total costs of enterprises in the sale of goods, products, works, services, rubles; C_{sumb} – total costs of enterprises in the sale of goods, products, works, services in the basic version of modeling rubles;

$\sum_{i=1}^{n} V_i$ – the total volume of sales of goods, products, works, services by enterprises. units; n – the number of divisions of the enterprise, the volume of sales of goods, products, works, services of which is taken into account in the distribution of conditionally fixed costs of the enterprise; FR – the financial result of enterprises from the sale of goods, products, works, services, rubles; FR_b – the financial result of enterprises from the sale of goods, products, works, services in the basic version of modelling, rubles; ΔW – the increase in wages compared with the basic version of modelling, rubles; ΔFR – the increase in financial results compared with the basic version of modelling, rubles; ΔR – the increase in income of enterprises compared to the basic version of modelling, rubles; ΔC_{var} – the increase in conditionally variable costs of enterprises in the sale of goods, products, works, services compared to the basic version of the simulation, rubles; T_{inc} - income tax rate, %; T_{val} – the rate of value added tax (VAT), %; T_{profit} – profit tax rate, %; ΔT – the increase in total tax deductions compared to the basic version of the simulation, rubles; $Credit$ – the amount of credit funds needed for the development of the enterprise, rubles; ω_{var} – the share of conditionally variable costs in the structure of the cost of goods sold, products, works, services; ω_{fix} – the share of conditionally fixed costs in the structure of the cost of goods, products, works, services sold).

3 Results

The results of modeling using the developed economic and mathematical model (1)–(9) are presented in Table. 1. In column 1 gives the number of the modeling option, and column 2 of Table 1 presents the modeling options by year. The first simulation variant corresponding to the first line of Table 1 *is the basic one (2023), which specifies the values of the simulated parameters.* It is assumed that the revenue at the Elkon Mining and Metallurgical Combine in 2023 will amount to 6 billion rubles. With the number of 600 employees, the average monthly revenue per employee will be 6 billion rubles.: 600 people: 12 months a year = 833 333rub. (row 1, column 3 Table 1). When modeling, the revenue provided by the sovereign issue is growing progressively: in 2024, relative to the basic modeling option (2023), by 3%, in 2025, relative to 2024, by 4%, and so on until the last line (11th modeling option) corresponding to 2033 that is, every year the revenue growth rate increases by 1% compared to the previous year. This means that in 2033, compared to 2032, revenue will grow by 12%.

The average monthly cost of goods, products, works, and services sold per employee will amount to 800,000 rubles in 2023 (column 4) (Tables 2, 3, and 4).

Table 1. Modelling results.

Option number	Year	Average monthly revenue of enterprises per employee, rubles	The average monthly cost of goods, products, works, services sold per employee, taking into account the progressive labor incentive system, rubles	The share of conditionally fixed costs in the cost structure	The share of conditionally variable costs in the cost structure	Conditionally fixed costs, rubles	Conditionally variable costs, rubles	Growth of the financial result depending on revenue growth and cost reduction, rubles
1	2	3	4	5	6	7	8	9
1	2023	833,333.00	800,000.00	93.00%	7.00%	744,000.00	56,000.00	0.00
2	2024	858,332.99	824,000.00	92.81%	7.19%	744,000.00	57,680.00	22,320.00
3	2025	892,666.31	856,960.00	92.54%	7.46%	744,000.00	59,987.20	52,972.80
4	2026	937,299.63	899,808.00	92.19%	7.81%	744,000.00	62,986.56	92,821.44
5	2027	993,537.60	953,796.48	91.77%	8.23%	744,000.00	66,765.75	143,030.73
6	2028	1,063,085.23	1,020,562.23	91.24%	8.76%	744,000.00	71,439.36	205,122.88
7	2029	1,148,132.05	1,102,207.21	90.60%	9.40%	744,000.00	77,154.50	281,052.71
8	2030	1,251,463.94	1,201,405.86	89.84%	10.16%	744,000.00	84,098.41	373,307.45
9	2031	1,376,610.33	1,321,546.45	88.94%	11.06%	744,000.00	92,508.25	485,038.20
10	2032	1,528,037.47	1,466,916.56	87.87%	12.13%	744,000.00	102,684.16	620,232.40
11	2033	1,711,401.96	1,642,946.54	86.61%	13.39%	744,000.00	115,006.26	783,940.29

In the production and sale of goods, products, works, services, the total costs are divided into conditionally constant, presented in column 7 of Table 1 (744,000.00 rubles.), those that do not depend on the volume of production and sales, and conditionally variable, shown in column 8 of Table 1 (56,000.00 rubles.). In the basic version of modeling, the share of conditionally fixed costs in the structure of the cost of goods sold, products, works, services is 93%, and the share of conditionally variable costs in the structure of the cost of goods sold, products, works, services is 7% (see the first row, columns 5 and 6 of Table 1). Thus, with the growth of revenue, the unit cost is automatically reduced by reducing the share of conditionally fixed costs per unit of production. This means that the average monthly cost of goods, products, works, services sold per

Table 2. Modelling results (*continue*).

The average monthly nominal accrued salary of employees of the enterprise mining ore, taking into account the growth of the average revenue of the enterprise, rubles	Percentage of deductions for salary increases	Deductions for salary increases, rubles	Average monthly nominal accrued salary, taking into account the progressive labor incentive system, rubles	Wage growth index	Percentage of deductions for salary increase of employees of the enterprise providing ore extraction	The average monthly nominal accrued salary of employees of the enterprise providing ore extraction, taking into account the growth of the average revenue of the enterprise, rubles	Percentage of contributions to the development fund	Increase in contributions to the development fund, rubles
10	11	12	13	14	15	16	17	18
100,000.00	12.00%	0.00	100,000.00	1.000	50.00%	416,666.50	38.00%	0.00
103,000.00	13.95%	16,740.00	119,740.00	1.197	48.00%	411,999.84	38.05%	4,464.00
107,120.00	16.45%	39,729.60	146,849.60	1.468	46.00%	410,626.50	37.55%	10,594.56
112,476.00	19.43%	69,616.08	182,092.08	1.821	44,00%	412,411.84	36.57%	18,564.29
119,224.56	22.80%	107,273.04	226,497.60	2.265	42.00%	417,285.79	35.20%	28,606.15
127,570.28	26.47%	153,842.16	281,412.44	2.814	40.00%	425,234.09	33.53%	41,024.58
137,775.90	30.36%	210,789.53	348,565.43	3.486	38.00%	436,290.18	31.64%	56,210.54
150,175.73	34.37%	279,980.59	430,156.32	4.302	36.00%	450,527.02	29.63%	74,661.49
165,193.31	38.43%	363,778.65	528,971.95	5.290	34.00%	468,047.51	27.57%	97,007.64
183,364.57	42.44%	465,174.30	648,538.87	6.485	32.00%	488,971.99	25.56%	124,046.48
205,368.32	46.36%	587,955.21	793,323.53	7.933	30.00%	513,420.59	23.64%	156,788.06

Table 3. Modelling results (*continue*).

Average monthly contributions to the development fund, rubles	Income tax rate	Profit tax rate	VAT rate	Financial result, rubles	Income tax increase per month per employee, rubles	Increase in income tax per month per employee, rubles	Increase in VAT deductions per month per employee, rubles	The increase in income tax, income tax and VAT deductions per month per employee, rubles
19	20	21	22	23	24	25	26	27
33,333.00	13.00%	20.00%	20.00%	33,333.00	0.00	0.00	0.00	0.00
37,797.00	13.00%	20.00%	20.00%	34,332.99	2,566.20	200.00	4,664.00	7,430.20
43,927.56	13.00%	20.00%	20.00%	35,706.31	6,090.45	474.66	11,069.22	17,634.33
51,897.29	13.00%	20.00%	20.00%	37,491.63	10,671.97	831.73	19,396.01	30,899.71
61,939.15	13.00%	20.00%	20.00%	39,741.12	16,444.69	1,281.62	29,887.77	47,614.08
74,357.58	13.00%	20.00%	20.00%	42,523.00	23,583.62	1,838.00	42,862.58	68,284.19
89,543.54	13.00%	20.00%	20.00%	45,924.84	32,313.51	2,518.37	58,728.91	93,560.78
107,994.49	13.00%	20.00%	20.00%	50,058.08	42,920.32	3,345.02	78,006.51	124,271.84
130,340.64	13.00%	20.00%	20.00%	55,063.88	55,766.35	4,346.18	101,353.82	161,466.35
157,379.48	13.00%	20.00%	20.00%	61,120.91	71,310.05	5,557.58	129,604.06	206,471.70
190,121.06	13.00%	20.00%	20.00%	68,455.42	90,132.06	7,024.48	163,812.54	260,969.08

employee, determined by the sum of conditionally fixed and conditionally variable costs (columns 7 and 8 of Table 1), significantly less than the same value of column 4, and the resulting difference can be directed to an increase in wages (a progressive labor incentive system) and contributions to the development fund. Cost reduction by reducing the share of conditionally fixed costs per unit of production allows the introduction of progressive remuneration of workers (column 13 of Table 1), the essence of which is that with the growth of revenue, the percentage of deductions from revenue for wages increases from 12.00% to 46.36% *due to the growth of the financial result due to revenue growth and cost reduction, which is reflected in column 9.* The share of conditionally fixed costs in the structure of the cost of goods, products, works sold, service ω_{fix} is determined by the formula (7) of the economic and mathematical model (1)–(9), and the share of conditionally variable costs in the structure of the cost of goods sold, products, works, services

Table 4. Modelling results (*ending*).

The increase in income tax per year from all citizens working at the enterprise, rubles	The increase in income tax per year from all citizens working at the enterprise, rubles	The increase in VAT deductions per year from all citizens working at the enterprise, rubles	The increase in income tax, income tax and VAT deductions from all citizens working at the enterprise, per year, rubles	The amount of credit funds for the development of the enterprise per month per employee, rubles	The amount of credit funds for the development of the enterprise per month per employee, minus the increase in deductions to the development fund, rubles	The amount of credit funds for all working citizens of the enterprise for the development of the enterprise per year, minus the increase in deductions to the development fund, rubles
28	**29**	**30**	**31**	**32**	**33**	**34**
0.00	0.00	0.00	0.00	0.00	0.00	0.00
187,953,329.03	14,648,230.81	341,600,011.96	544,201,571.81	1,680.00	-2,784.00	-203,905,411.90
446,075,900.90	34,765,134.45	810,730,695.06	1,291,571,730.42	3,987.20	-6,607.36	-483,935,510.92
781,635,244.34	60,917,109.19	1,420,600,583.09	2,263,152,936.61	6,986.56	-11,577.73	-847,974,639.64
1,204,440,017.06	93,868,597.36	2,189,036,642.00	3,487,345,256.42	10,765.75	-17,840.39	-1,306,663,941.82
1,727,308,586.00	134,618,604.39	3,139,335,901.52	5,001,263,091.92	15,439.36	-25,585.22	-1,873,909,712.19
2,366,702,150.30	184,450,041.56	4,301,416,138.88	6,852,568,330.75	21,154.50	-35,056.04	-2,567,570,254.24
3,143,565,330.93	244,995,237.73	5,713,343,627.28	9,101,904,195.93	28,098.41	-46,563.08	-3,410,367,812.84
4,084,432,960.80	318,322,197.53	7,423,344,696.55	11,826,099,854.88	36,508.25	-60,499.39	-4,431,089,300.47
5,222,882,792.94	407,047,818.88	9,492,445,990.37	15,122,376,602.20	46,684.16	-77,362.32	-5,666,162,300.50
6,601,442,044.22	514,486,480.38	11,997,939,557.08	19,113,868,081.68	59,006.26	-97,781.80	-7,161,723,424.17

ω_{var} according to formula (8). The value of conditionally variable costs is proportional to the volume of products produced and sold, goods, works, services.

As follows from the comparison of the data presented in columns 10 and 13 of Table 1, the use of a progressive VAT labor incentive system described by formula (1) and acting as the objective function of the economic and mathematical model (1)–(9) leads to a significantly faster increase in average monthly wages compared to the data in column 10 of Table 1, namely: 7.933 times compared with the baseline a variant of modeling, which can be seen in the last row of column 14 of Table 1, where the wage growth index is shown.

The financial result shown in column 23 of Table 1 is calculated according to the formula (6) of the economic and mathematical model (1)–(9) and is equal to the difference in the values presented in columns 3 and 4. The growth of the financial result depends on the growth in the volume of manufactured and sold products, goods, works, services and a decrease in The cost is shown in column 9 of Table 1, calculated by the formula (4) of the economic and mathematical model (1)–(9) and represents a decrease in the values of column 4 by the values presented in columns 7 and 8.

Percentage of deductions for salary increases (column 11 of Table 1) was calculated by the formula (3) of the economic and mathematical model (1)–(9). In accordance with formulas (1) and (2) of the economic-mathematical model (1)–(9), deductions for wage increases and deductions to the development fund were calculated, presented in columns 12 and 18, respectively. At the same time, the coefficient of redistribution of the increase in the financial result between working citizens and owners of enterprises (ξ) is assumed to be 0.75, which means that 75% of the increase in the financial result of enterprises from the sale of goods, products, works, services is directed to increase the wages of employees working at these enterprises, and 25% goes to the enterprise development fund. This distribution of the financial result is taken from the ratio of salary and financial result in the basic version (100,000 rubles: 33,333 rubles = 3). We will show an algorithm for calculating deductions for salary increases and to the enterprise development fund using the example of the second row of columns 12 and 18 of Table 1. According to formula (6), the financial result is 858,332.99 rubles (the average monthly revenue of the enterprise per employee, see row 2, column 3 of Table 1) – 744,000.00 rubles (conditionally fixed costs, see column 7 of Table 1) – 57 680,00 rubles (conditionally variable costs, see column 8 of Table 1) = 56,652.99 rubles. The growth of the financial result, depending on revenue growth and cost reduction, is equal to 22,320.00 rubles (see the second row, column 9 of Table 1) = 56,6652.99 rubles – 34,332.99 rubles (financial result in the second version of the simulation, see row 2, column 23 of Table 1). Then for the second row of column 12 of Table 1, deductions for salary increases according to formula (1) are equal to 16,740.00 rubles = 22,320.00 rubles (the growth of the financial result depending on revenue growth and cost reduction, see row 2, column 9 of Table 1) • 0.75 (the coefficient of redistribution of the increase in financial results between working citizens and owners of enterprises), and the increase in contributions to the development fund (see the second row, column 18 of Table 1) is 4,464.00 rubles = 22,320.00 rubles (the growth of the financial result depending on revenue growth and cost reduction, see row 2, column 9 of Table 1) • (1 – 0.75 (the coefficient of redistribution of the increase in the financial result between working citizens and owners of enterprises) • 0.8 (correction of the result due to the need to take into account the income tax rate, 20%, see formula (2)). Similarly for other rows of columns 12 and 18 of Table 1. It is worth noting that for the 10th modeling option, the increase in contributions to the development fund in absolute terms amounted to 156,788.06 rubles per employee (see the last row of column 18 of Table 1).

Column 19 of Table 1 shows the average monthly contributions to the development fund, determined by adding an increase in contributions to the development fund to the base amount of the financial result (see formula (2)). So, for the second row of column 19 of Table 1 value 37,797.00 rubles = 33,333.00 rubles (see the first row, column 19 of Table 1) + 4,464.00 rubles (see the second row, column 18 of Table 1). Similarly for the remaining rows of column 19 of Table 1.

The increase in income tax per month per one working (column 24 of the Table 1) is determined by multiplying the wage increment shown in column 13 relative to the basic modeling option by the income tax rate (13%, column 20). For the second modeling option (2024), the increase in income tax per month per citizen working at the enterprise is 2,566.20 rubles = 19,740.00 rubles • 0.13 (income tax rate), where 19,740.00 rubles – this is an increase in wages compared to the basic version of the simulation (ΔW in formula (5) of the economic and mathematical model (1)–(9)), which is calculated as follows: 19,740.00 rubles = 119,740.00 rubles (see the second row, column 13 of Table 1) – 100,000 rubles (the basic version of the simulation, column 13 of Table 1). Similarly for the remaining rows of column 24 of Table 1.

The increase in income tax per month per employee (column 25 of Table 1) is determined by multiplying the increase in the financial result shown in column 23 relative to the basic modeling option by the income tax rate (20%, column 21). For the second modeling option (2024), the increase in profit tax per month per citizen working at the enterprise is 200.00 rubles = 999.99 rubles • 0.20 (income tax rate), where 999.99 rubles – this is an increase in the financial result compared to the basic version of the simulation (ΔFR in formula (5) of the economic and mathematical model (1)–(9)), which is calculated as follows: 999.99 rubles = 34,332.99 rubles (see the second row, column 23 of Table 1) – 33,333.00 rubles (basic modeling option, column 23). Similarly for the remaining rows of column 25 of Table 1.

Finally, the increase in VAT deductions per month per employee (column 26 of Table 1) is determined by multiplying the increase in income of enterprises from the sale of goods, products, works, services shown in column 3, minus the increase in conditionally variable costs (column 8) relative to the basic modeling option by the VAT rate (20%, column 22). For the second modeling option (2024), the increase in VAT deductions per month per citizen working at the enterprise is equal to 4,664.00 rubles = 23,320.00 rubles • 0.20 (VAT rate), where 23,320.00 rubles is an increase in the income of enterprises from the sale of goods, products, works, services (ΔR in formula (5)) minus the increase in conditionally variable costs (ΔC_{var} in formula (5)) compared to the basic modeling option, which is calculated as follows: 23,320.00 rubles = 858,332.99 rubles (see the second row, column 3 of Table 1) – 833,333.00 rubles (basic modeling option, column 3) – 57,680.00 rubles (see the first row, column 8 of Table 1) + 56,000.00 rubles (basic modeling option, column 8). Similarly for the remaining rows of column 26 of Table 1.

The values presented in column 27 are the sum of the values of the corresponding rows of columns 24–26.

600 people work at this enterprise, this value is used in calculating the data presented in columns 28–31 and 34 of Table 1. So, 187,953,329.03 rubles (the increase in income tax per year from all citizens working at the enterprise for 2024, see the second row, column 28 of Table 1) = 2,566.20 rubles (increase in income tax per month per employee, see the second row, column 24 of Table 1) • 600 people • 12 (the number of months in a year). Similarly for all other rows of columns 28–31 and 34.

The amount of credit funds for the development of the enterprise per month per employee is determined by the increase in the total costs of enterprises in the sale of goods, products, works, services relative to the basic modeling option (formula (9) of the economic and mathematical model (1)–(9)). For example, for the second row of column 32, the value is 1,680.00 rubles = 57,680.00 rubles (second row, column 8 of Table 1) – 56,000.00 rubles (the basic version of the simulation, see column 8), and the data in column 33 is a decrease in the values of column 32 by the amount of the increase in contributions to the enterprise development fund. In particular, for the second row of column 33, the value -2,784.00 rubles = 1,680.00 rubles (second row of column 32) – 4,464.00 rubles (increase in contributions to the development fund, second row, column 18 of Table 1).

The amount of credit funds for all working citizens of the enterprise for the development of the enterprise per year, minus the increase in contributions to the development fund (see column 34 of Table. 1) is equal to the product of the amount of credit funds per month per citizen working at the enterprise, minus the increase in deductions to the development fund (column 33) by the average number of workers at the enterprise (600 people) and by the number of months per year. For example, for the second row of column 34 of Table 1 value -203,905,411.90 rubles = -2 784.00 rubles (see the second row of column 33 of Table 1) • 600 people • 12 (number of months in a year). Similarly for the remaining rows of column 34 of Table 1.

4 Discussion

The analysis of the results obtained using economic and mathematical modeling (Table 1) allows us to draw the following conclusions:

Due to the sovereign issue and social financial technologies, with quite real revenue growth rates, tax revenues from the enterprise in question for the third year of the sovereign issue (Table 1, row 3, column 31) amounted to 1,291,571,730.42 rubles.
The company's revenue per employee has increased by 2.05 times over 10 years (Table 1, Column 3), wages by 7.933 times (Table 1, column 14).
Deductions for the development of the enterprise over 10 years have increased from one employee by 5.70 times (Table 1, column 19), and amounted to 1,368,871,632.00 rubles for the entire workforce (600 people) over the year = (190,121.06 • 600 • 12 = 1,368,871,632.00 rubles).

5 Conclusion

The state, providing interest-free loans to the enterprise and completely writing off the loan debt (the body of the loan), receives 3 times more funds to the budget than it invests in the enterprise.

It was noted above that with a sovereign issue in China, if an enterprise fulfills the planned indicators within three years, the state writes off 30% of the loan body to it, and another 20% a year later [9, 10].

This article shows that sovereign lending with social financing technologies allows for 10 years to increase income tax receipts three times, value added taxes – 2 times, income tax – 7.8 times, which with a margin will exceed the entire body of the loan. *In other words, the company, in case of effective work, will not need to repay the loan.*

In contrast to the Chinese version, the economic and mathematical model proposed in the article allows to determine individually for each enterprise and for the years of sovereign financing, taking into account the growth of tax revenues, the specific terms and volumes of write-off by the state of the loan body.

References

1. Sokolov, E.V., Kostyrin, E.V., Rudnev, K.V.: Social financial technologies for the development of enterprises and the economy of Russia. Soft Meas. Calculations **9**, 35–46 (2021)
2. Shestakova, E.: Social protection programs before and during the pandemic: old and new challenges. Soc. Econ. **1**, 81–99 (2021)
3. Baeten, R., Spasova, S., Vanherche B., Coster, S.: Inequalities in access to healthcare. a study of national policies. Brussels: European Commission (2018)
4. Omelchenko, I.N., Gertsik, Y.G.: Opportunities and challenges of initiative-innovative cluster structures within large-scale systems of the medical industry and healthcare. In: Proceedings of 2021 14th International Conference Management of Large-Scale System Development, MLSD 2021 (2021)
5. Social Fund of Russia. Open data of the Pension Fund of Russia until 2023. http://www.sfr.gov.ru/opendata/pfr_opendata. Accessed 14 Nov 2022

6. Federal Law "On the Subsistence Minimum in the Russian Federation" dated 24.10.1997 No 134-FZ. ConsultantPlus: reference right system: ofic. Website, Company "ConsultantPlus". http://www.consultant.ru/data. Accessed 14 Nov 2022
7. Ganiev, E., Rahmatov, A.: Improving the institutions of the continuing education system and an economical mathematical model of optimal placement. Theoret. Appl. Sci. **12**(104), 587–589 (2021)
8. Pereira, G.: and. In: Schweiger, G. (ed.) Poverty, Inequality and the Critical Theory of Recognition. PP, vol. 3, pp. 83–106. Springer, Cham (2020). https://doi.org/10.1007/978-3-030-45795-2_4
9. Zhu, T.S.: How does China choose the path of global governance reform? Bulletin of International Organizations: Education, Science, New Economy, vol. 3(15), pp. 248–281 (2020)
10. Gunchinsharav, B., Velichkin, S.V.: The impact of the special military operation conducted in Ukraine on Russia's relations with East Asian countries. Diplomatic Serv. **4**, 250–264 (2022)

Development of a Software Product for Calculating the Trajectory of the Socio-Economic Development of the Region

Ekaterina S. Chernova(✉) and Vladislav V. Kalinin

Kemerovo State University, Kemerovo, Kemerovo Region, Russia
elvangie@mail.ru

Abstract. The article is devoted to the construction of a mathematical model, the development of a computational algorithm and a software product for the formation of a strategy and trajectory of the socio-economic development of the region (on the example of the Kemerovo region - Kuzbass). The methods of mathematical statistics, the theory of optimal discrete processes, computational mathematics, and simulation are used as research methods. The phase states in the proposed model are the indicators for six sustainable development goals adopted by the UN in 2015 in the "2030 Agenda for Sustainable Development": "Eradication of poverty", "Eradication of hunger", "Good health", "Decent work and economic growth", "Industrialization, innovation and infrastructure", "Peace, justice and effective institutions". As a result of the study, an algorithm has been implemented that makes it possible to determine the predicted values of the listed indicators based on the distribution of the budget for eleven items of expenditure. The developed software product allows you to vary the values of the shares of budget expenditures allocated to specified areas of financing, thus considering various scenarios for the development of the region, and choosing the most appropriate option for a socio-economic development strategy from the point of view of decision makers. The proposed methodology for building a model and developing appropriate software can be used for any region, taking into account the specifics of its development.

Keywords: Sustainable Development · Mathematical Modeling · Software Product · Regional Development Strategy

1 Introduction

In various regions of the world and individual countries, there is a wide range of socio-economic problems due to geographical location, natural and climatic factors, demographic situation and other conditions. In 2015, the UN member states formulated seventeen sustainable development goals [1] aimed at solving global problems related to eradicating poverty, ensuring environmental protection, improving the quality of life and improving well-being for the world's population. However, to achieve these goals

© The Author(s), under exclusive license to Springer Nature Switzerland AG 2023
A. Gibadullin (Ed.): DITEM 2022, LNNS 683, pp. 77–89, 2023.
https://doi.org/10.1007/978-3-031-30926-7_8

in each individual region, it is necessary to develop its own approach that would take into account the problems and characteristics of the territory under study, its strengths and weaknesses. One way to solve this problem can be the use of mathematical and computer modeling, which will allow us to analyze the existing problems of regional development from a quantitative point of view.

In the last seven years (after the adoption of the "2030 Agenda for Sustainable Development"), research in the field of mathematical and computer modeling of the problems of sustainable development of regions has been concentrated in several areas.

The first group of works [2–10] is related to the discussion, formalization, forecasting of indicators and the index of sustainable development. In particular, in [2] various systems of indicators and principles of sustainable development are considered, their advantages and disadvantages are described. In [3], options for constructing a sustainable development index are studied, including using the system dynamics method. In [4] and [5], the goals of sustainable development and the need to build reasonable indicators for each of them are discussed. In [6], linear econometric equations for the sustainability of the region's development are constructed. In [7], the innovative rating of regions is studied as a condition for achieving a qualitatively new state of the economy. In [8], an index of regional sustainability was developed through the level of maturity of entrepreneurial ecosystems. In [9], an index of sustainable urbanization was developed. In [10], the relationship between the seventeen goals of sustainable development is determined on the basis of a correlation analysis.

Most of the works are devoted to research in the framework of specific areas of sustainable development of a particular region - social problems, education, protection of natural resources, etc. For this purpose, econometric methods [11], neural networks [12], multicriteria decision-making methods [13] are used. There are works devoted to the construction of optimization problems for the analysis of individual indicators for sustainable development goals [14] or for the analysis of the problem of food waste [15], the application of graph theory to study the partnership of regions in the field of sustainable development [16] and the theory of complex networks for analysis daily migration [17], the construction of nonlinear differential equations for the study of urban systems [18], the development of a mathematical formula for harmony to ensure the sustainable development of man and mankind [19]. Regression models are widely used: issues of sustainable human resource management [20], the impact of sustainable development indicators on economic prosperity [21] are being studied. An important role in the study of the economic problems of cities [22], regions [23], the sustainability of the transport industry [24] is played by the method of system dynamics.

In addition, the risks of using mathematical decision-making methods in some areas of sustainable development are analyzed [25], the spatio-temporal dynamics of fires and carbon emissions in certain areas are studied [26], hydrological models are built to solve problems of sustainable development [27], educational models of higher educational institutions for the integration of the principles and practices of sustainable development [28].

There are practically no comprehensive studies aimed at studying the possibilities of building a socio-economic strategy within the framework of sustainable development goals, taking into account the specifics of the region. Thus, the purpose of this article is to develop a methodology for constructing a mathematical model and implementing its software implementation to study the impact of the distribution of financial resources in the region on achieving sustainable development goals.

2 Materials and Methods

To achieve the goal set in the work, methods of mathematical modeling, regression analysis, the theory of optimal discrete processes, computational mathematics, and simulation modeling are used.

Let us consider step by step the process of constructing a mathematical model of the socio-economic development of the region.

2.1 Building a Mathematical Model

Definition of Phase Variables. At the first stage of the study, the following six sustainable development goals were selected for analysis from the list formulated in the 2030 Agenda for Sustainable Development: “Eradicating poverty”, “Zero hunger”, “Good health”, “Decent work and economic growth”, “Industrialization, innovation and infrastructure”, “Peace, justice and effective institutions”. For each of the goals, as an example, one quantitative indicator was determined that characterizes the degree of its achievement. Such indicators play the role of phase variables in the developed mathematical model. The list of goals, relevant indicators and designations adopted for them is given in Table 1.

Table 1. Sustainable development goals and indicators of their achievement.

Sustainable Development Goal	Indicator (phase variable of the model)	Designation
“Eradication of Poverty”	The share of the population with cash incomes below the subsistence level	x_1
“Hunger Elimination”	Agricultural products	x_2
“Good health”	Died from all diseases	x_3
“Decent Work and Economic Growth”	Gross regional product	x_4
“Industrialization, Innovation and Infrastructure”	Number of enterprises and organizations	x_5
“Peace, Justice and Effective Institutions”	Number of registered crimes	x_6

Definition of Control Parameters. At the second stage of the study, the shares of the region’s budget expenditures for eleven items were selected as parameters for managing the indicators considered in Table 1. The list of control parameters and corresponding designations is given in Table 2.

Table 2. Control parameters of the model.

Budget expenditure items (control parameters)	Designation
"National Issues"	u_1
"National Security and Law Enforcement"	u_2
"National economy"	u_3
"Department of Housing and Utilities"	u_4
"Environmental Protection"	u_5
"Education"	u_6
"Culture and Cinematography"	u_7
"Health"	u_8
"Social politics"	u_9
"Physical Culture and sport"	u_{10}
"Mass media"	u_{11}

Construction of Equations of Dynamics. At the third stage of the study, using the Statistica software package, equations for the dynamics of the indicators under consideration are constructed. Modeling is carried out on the example of statistical data for the Kemerovo region - Kuzbass from 2007 to 2021, which can be found on the website of the Federal State Statistics Service [29]. Autoregression models were used to construct dynamic equations. Indicators for six sustainable development goals were selected as dependent variables, and the shares of budget allocations for various items of expenditure were selected as factors. Stepwise regression with inclusion was used to select the variables that make the greatest contribution to the indicator variation. Both linear and logarithmic equations were built, and based on the obtained coefficient of determination, the most acceptable model was chosen, containing only significant variables in the regression equation.

As a result, for the Kemerovo region - Kuzbass, the following equations for the dynamics of the indicators under consideration were obtained, where t denotes the current moment in time (year), and T denotes the planning horizon.

$$x_1(t+1) = 0.249x_1(t) - 0.830ln(u_2(t)) - 0.280ln(u_4(t)), \quad (1)$$

$$x_2(t+1) = 0.957x_2(t) + 0.080ln(u_7(t)), \quad (2)$$

$$x_3(t+1) = 0.661x_3(t) + 0.343ln(u_1(t)), \quad (3)$$

$$x_4(t+1) = 0.903x_4(t) - 0.210u_5(t) + 0.427u_9(t) - 0.360u_{10}(t) + 0.477u_{11}(t), \quad (4)$$

$$x_5(t+1) = 0.886x_5(t) + 0.390u_1(t), \quad (5)$$

$$x_6(t+1) = 0.841x_6(t) + 1.030u_3(t) + 0.371u_4(t)$$
$$+1.370u_6(t) - 2.300u_9(t) - 1.100u_{11}(t), t = 0, 1, \ldots, T - 1. \quad (6)$$

Determination of Initial States and Restrictions. At the fourth stage of the study, the obtained equations of the form (1)–(6) act as equations of the dynamics of the model for managing the development of the region. They are supplemented by the state of the region at the moment of time, which in the calculations will be taken as the initial one, as well as restrictions on control parameters and phase restrictions. The first of the model's limitations should reflect the fact that the sum of the budget expenditure shares for all the items under consideration at each point in time should not exceed 1. The second group of restrictions is the conditions for the minimum and maximum amounts of funding for each of the expenditure items. These values can be varied, thus obtaining simulation results for different simulation experiments. In the example given in this article, when determining the limiting volumes of funding, the maximum and minimum values of the shares of budget expenditures by items from the statistical data for the Kemerovo region - Kuzbass from 2007 to 2021 were found. The considered funding interval was expanded by adding to the maximum values the value of 0.01 and the subtraction from the minimum values of the value 0.01. This procedure is used to expand the boundaries of the use of budgetary funds in order to obtain a more flexible model. The third group of restrictions are non-negativity restrictions on the phase variables of the model, which follow from the meaningful meaning of the indicators under consideration. Finally, the fourth constraint applies only to the first phase variable, which is measured as a percentage and therefore must be bounded from above by 100.

The general view of the initial states of the region can be represented as follows:

$$x_i(0) = x_i^0, \ i = 1, \ldots, 6. \quad (7)$$

The first constraint of the model is as shown below.

$$u_1(t) + u_2(t) + \ldots + u_{11}(t) \leq 1, \ t = 0, 1, \ldots, T. \quad (8)$$

The second group of restrictions has the following form:

$$0.0128 \leq u_1(t) \leq 0.1444, \ 0.0010 \leq u_2(t) \leq 0.1170, \ 0.0971 \leq u_3(t) \leq 0.3578, \quad (9)$$

$$0.0128 \leq u_4(t) \leq 0.0690, \ 0.0010 \leq u_5(t) \leq 0.0122, \ 0.1122 \leq u_6(t) \leq 0.3286, \quad (10)$$

$$0.0013 \leq u_7(t) \leq 0.0348, \ 0.0754 \leq u_8(t) \leq 0.2741, \ 0.0504 \leq u_9(t) \leq 0.4026, \quad (11)$$

$$0.0010 \leq u_{10}(t) \leq 0.0261, \ 0.0010 \leq u_{11}(t) \leq 0.0137, \ t = 0, 1, \ldots, T. \quad (12)$$

Next, we present the phase constraints of the model:

$$x_i(t) \geq 0, i = 1, \ldots, 6, \ t = 0, 1, \ldots, T, \quad (13)$$

$$x_1(t) \leq 100, \ t = 0, 1, \ldots, T. \quad (14)$$

Model Conclusions. Problem (1)–(14) is a discrete time control problem. It is known that problems of this type are almost always solvable, and often there are many controls that solve one problem. Therefore, more complex optimal control problems are posed, in which the quality functional is additionally specified. However, within the framework of this work, such functionality is not considered, since the goal was to create a software product that provides the ability to analyze user-defined scenarios for the behavior of a regional system using various simulation experiments.

2.2 Algorithm for Solving the Problem

To solve the problem (1)–(14), the following algorithm was formulated, which makes it possible to automate the process of calculating the trajectory of the socio-economic development of the region.

- Collection of statistical data on the initial states of the indicators listed in Table 1.
- Choice of management parameters, that is, the share of budget expenditures directed to the eleven areas of financing listed in Table 2.
- Checking the admissibility of the control constructed at step 2, that is, checking the fulfillment of conditions (8)–(12). If any of the conditions are violated, you must return to step 2.
- Building with the help of a given control of the corresponding trajectory of the socio-economic development of the region, that is, determining the indicators of sustainable development of the system using phase Eqs. (1)–(6). Calculation of indicators is carried out by successive substitution of the selected control parameters into relations (1)–(6).
- Checking the constructed trajectory for admissibility using phase constraints (13)–(14). If any of the conditions are not met, you must return to step 2.

The trajectory found at step 5 will be the trajectory of the socio-economic development of the region, corresponding to the control parameters specified at step 2. Any function u defined over the entire time interval $[0, T]$ and at each moment t taking a value from a given set determined by relations (8)–(12) will be called a strategy for the socio-economic development of the region.

2.3 Software Product Development

Consider the issue of developing a software product that allows you to automate the process of calculating the trajectory of the socio-economic development of the region based on the available data.

Tools Used. The RAD Studio development environment and the C++ programming language were used to create the code and interface. The interface prototype includes the following components: Button - for entering data and performing calculations, Label - for creating element labels, StringGrid - for creating data tables, TextBox - for storing reference information.

Input and Output Data. The following list of inputs for the software product has been defined.

- Data for the six sustainable development indicators listed in Table 1 at the start of time.
- Maximum and minimum shares of budget expenditures for each of the eleven items listed in Table 2.
- Estimated shares of budget expenditures for each of the articles for the entire planning period under consideration.
- Coefficients for significant parameters of the equations of dynamics (1)–(6).

The consideration of the coefficients as input data expands the possibilities of the program and allows it to be used both in further studies of Kuzbass under the conditions of changed development trends, and for other regions for which the form of functions on the right-hand sides of the dynamics equations coincides with their form in (1)–(6), but the coefficients at the control parameters are different. These coefficients should be determined by specialists in the field of statistical analysis at the preliminary stage of the study of the region.

The output data of the program are the calculated values of indicators for sustainable development goals.

Interface Prototype Development. The program interface includes four tables:

- Estimated shares of budget expenditures by item ("Shares of Budget Expenditures");
- Maximum and minimum values of budget expenditure shares by items (marked as "MAX" and "MIN", respectively);
- Coefficients for significant parameters from dynamics Eqs. (1)–(6) ("Coefficients for Significant Parameters");
- Calculated values of indicators for sustainable development goals ("Values of Indicators for Regional Development").

Buttons with functions are provided for each of the listed four tables:

- Data entry into the table from a file;
- Entering data into the table manually;
- Resetting the setpoints from the table.

The "Shares of Budget Expenditures" table has a "Sum" button to check the constraint (8). The "Values of Indicators for Regional Development" table has a "Calculation of Indicators" button, which, when clicked, performs calculations using dynamics Eqs. (1)–(6).

In the right part of the program window, there is also reference information on the symbols used.

The interface of the software product is shown in Fig. 1.

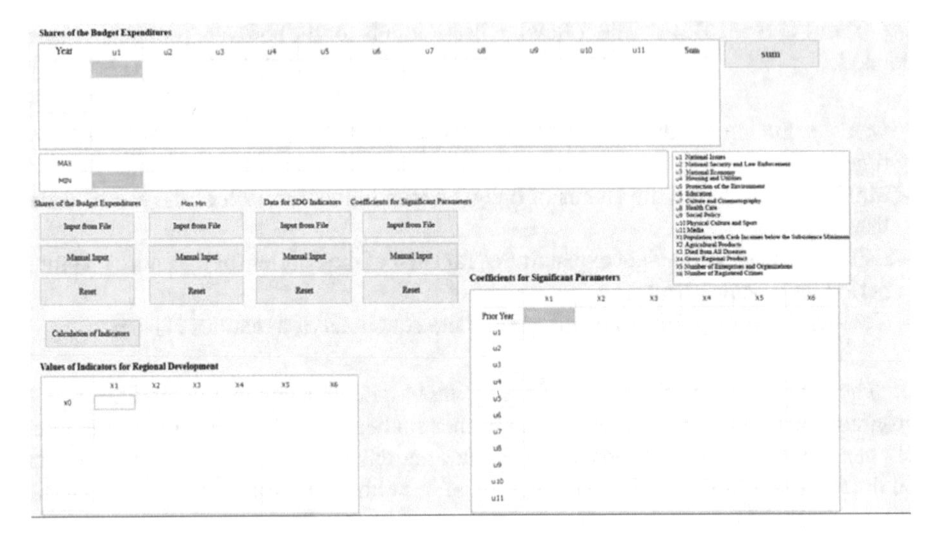

Fig. 1. View of interface of the software product.

3 Results

Let us give an example of the operation of the software product described in the previous section.

At the first step, we will fill in the table "Shares of Budget Expenditures". This can be done in either of the two suggested ways. We will consider 2018 as the starting point in time and enter data for a five-year period until 2022 in order to analyze the existing development trends in the region and compare the actual development trajectory with that built using a simulation experiment. As an example, we consider a scenario obtained as a result of one of the experiments, which allows reducing the share of the population with monetary incomes below the subsistence level. The corresponding input data are shown in Table 3.

Table 3. Shares of budget expenditures.

Year	u_1	u_2	u_3	u_4	u_5	u_6	u_7	u_8	u_9	u_{10}	u_{11}
2018	0.0387	0.1158	0.2611	0.0450	0.0034	0.2145	0.0086	0.1233	0.1616	0.0186	0.0038
2019	0.0321	0.0950	0.2504	0.0501	0.0037	0.1174	0.0105	0.0818	0.1430	0.0252	0.0043
2020	0.0530	0.1135	0.2643	0.0584	0.0018	0.1748	0.0243	0.1287	0.1450	0.0154	0.0063
2021	0.0480	0.0899	0.1181	0.0585	0.0090	0.1545	0.0143	0.2442	0.2580	0.0001	0.0048
2022	0.0416	0.0996	0.1343	0.0227	0.0081	0.1234	0.0143	0.2263	0.2305	0.0052	0.0119

Using the "Sum" button, the sum of the budget expenditure shares for the line is checked, which should not exceed 1.

At the second step, in the next table we will enter the maximum and minimum values of the shares of budget expenditures for each of the articles under consideration. The corresponding quantities are given in relations (9)–(12).

At the third step, in the table "Coefficients for Significant Parameters", we introduce the coefficients from relations (1)–(6), and the coefficients before $x_i(t)$, $i = 1,\ldots, 6$, from the corresponding dynamics equation are entered in the "Prior Year" line. In the absence of a control parameter in the equation of dynamics, a zero value is entered into a given cell.

At the fourth step, in the first row of the "Values of Indicators for Regional Development" table, we will enter the values of the regional development indicators at the initial moment of time. After clicking on the "Calculation of Indicators" button, the calculated values of the six indicators considered in the model for sustainable development goals will be displayed in the next rows of the table. The result of the program for the described scenario is shown in Fig. 2.

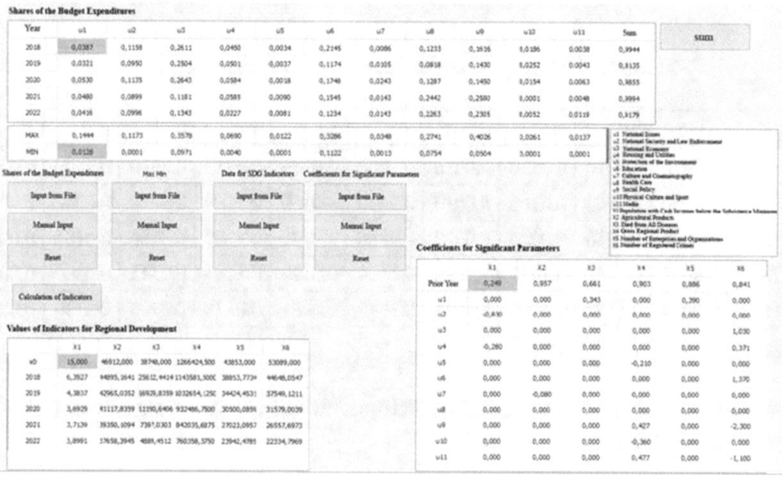

Fig. 2. An example of the software product.

The final values of indicators for sustainable development goals are presented in Table 4.

Table 4. Final values of indicators of regional development.

Year	x_1	x_2	x_3	x_4	x_5	x_6
2017	15.000	46 912.0000	38 748	1 266 424.5000	43 853	53 089
2018	6.3927	44 895.1641	25 612	1 143 581.5000	38 854	44 648
2019	4.3837	42 965.0352	16 930	1 032 654.1250	34 424	37 549
2020	3.6929	41 117.8359	11 191	932 486.7500	30 500	31 579
2021	3.7139	39 350.1094	7 397	824 035.6875	27 023	26 558
2022	3.8991	37 658.3945	4 889	760 358.3750	23 942	22 335

Let's compare the obtained results with the calculated values of indicators according to the scenario corresponding to the actual distribution of budgetary funds for the considered period of time (Table 5).

Table 5. The actual distribution of budget expenditures.

Year	u_1	u_2	u_3	u_4	u_5	u_6	u_7	u_8	u_9	u_{10}	u_{11}
2018	0.0240	0.0050	0.1410	0.0320	0.0010	0.3000	0.0150	0.0850	0.3930	0.0040	0.0010
2019	0.0270	0.0050	0.1560	0.0200	0.0010	0.2970	0.0150	0.1100	0.3620	0.0060	0.0005
2020	0.0230	0.0080	0.1710	0.0470	0.0010	0.2880	0.0170	0.1110	0.3180	0.0160	0.0010
2021	0.0182	0.0028	0.1241	0.1035	0.0029	0.2108	0.0114	0.1140	0.2511	0.0395	0.0004
2022	0.0201	0.0029	0.1198	0.1063	0.0035	0.1945	0.0084	0.0871	0.2746	0.0139	0.0001

The calculated values of the region's development indicators for a given distribution of budgetary funds are given in Table 6.

Table 6. Values of indicators of regional development in the actual distribution of budget expenditures.

Year	x_1	x_2	x_3	x_4	x_5	x_6
2017	15.0000	46 912.0000	38 748	1 266 424.5000	43 853	53 089
2018	9.0964	44 894.4480	25 611	1 143 581.4901	37 977	44 648
2019	7.7580	42 963.6508	16 928	1 032 654.2380	32 888	37 548
2020	6.7954	41 115.8878	11 188	932 486.9072	28 481	31 578
2021	7.2099	39 347.5466	7 394	842 035.7699	24 664	26 557
2022	7.2669	37 655.2202	4 886	760 358.4117	21 359	22 334

4 Discussion

Simulation experiments carried out using the developed software product demonstrate the following results. During the entire planning period, an increase in the share of budget allocations was made under the items of expenditure "National issues", "National security and law enforcement", "Environmental protection", "Mass media", and also within four years - under the item "National economy". A slight decrease in budget expenditures throughout the entire five-year period under review was made under the item "Education", and also within four years - under the item "Social Policy". Under the item "Housing and communal services" expenditures were increased in the first three years, and under the item "Culture and cinematography" - in the last three years. In the last three years, on the contrary, budget expenditures have been reduced under the item "Physical culture and sports". The greatest changes in the first three years have affected the articles "National Economy" and "Social Policy", in the last two years - the articles "Housing and Communal Services" and "Education".

Let us analyze the development trajectories of the region, built as a result of the implementation of the two described strategies, one of which corresponds to the actual one. The proposed scenario for changing budget expenditures in comparison with actual ones leads in the future to a decrease in the proportion of the population with monetary incomes below the subsistence level (by about 3%), while other development indicators are stable, there is a decrease in mortality and the number of registered crimes.

5 Conclusion

The conducted studies allow us to form a general methodology for constructing the trajectory of the socio-economic development of the region and give recommendations on the development of software products for studying regions with different development specifics. The introduction of applications for the analysis of regional development trajectories will help to explore various strategies for allocating budgetary funds and their impact on the achievement of sustainable development goals.

In this paper, only some examples of development scenarios are considered, and the results of calculations largely depend on the chosen goals and the introduced phase variables. The conducted study demonstrates the general methodology for implementing such calculations using the example of the Kemerovo region - Kuzbass, however, in order to obtain recommendations of great practical importance, further "expansion" of the model may be required: taking into account a larger number of sustainable development goals, as well as finalizing indicators for sustainable development goals in general.

References

1. 2030 Agenda for Sustainable Development. https://sdgs.un.org/2030agenda. Accessed 01 Feb 2023
2. Bolshakov, B.E., Shamaeva, E.F.: Sustainable development: yesterday - today - tomorrow. Measurement problem. Online Mag. Sci. Sci. **9**(4), 1–23 (2017)

3. Costanza, R., et al.: Modelling and measuring sustainable wellbeing in connection with the UN sustainable development goals. Ecol. Econ. **130**, 350–355 (2016)
4. Hák, T., Janoušková, S., Moldan, B.: Sustainable development goals: a need for relevant indicators. Ecol. Ind. **60**, 565–573 (2016)
5. Janoušková, S., Hák, T., Moldan, B.: Global SDGs assessments: helping or confusing indicators? Sustainability **10**(5), 1540 (2018)
6. Vdovin, S.M., Gus'kova, N.D., Neretina, E.A., Ivanova, I.A.: Region's sustainable development prediction on the basis of economic-mathematical modeling. Natl. Interests Prior. Secur. **12**(9), 18–27 (2016)
7. Chulkova, G., Vorobeva, E., Vorobev, O.: Activization of scientific and innovative sphere for the region sustainable development. In: SHS Web of Conferences, vol. 94, p. 02011 (2021)
8. Tolstykh, T., Gamidullaeva, L., Shmeleva, N., Woźniak, M., Vasin, S.: An assessment of regional sustainability via the maturity level of entrepreneurial ecosystems. J. Open Innov. Technol. Mark. Complex. **7**(1), 5 (2021)
9. Zhong, L., Li, X., Law, R., Sun, S.: Developing sustainable urbanization index: case of China. Sustainability (Switzerland) **12**(11), 4585 (2020)
10. Fonseca, L.M., Domingues, J.P., Dima, A.M.: Mapping the sustainable development goals relationships. Sustainability (Switzerland) **12**(8), 3359 (2020)
11. Su, A., He, W.: Exploring factors linked to the mathematics achievement of ethnic minority students in China for sustainable development: a multilevel modeling analysis. Sustainability **12**(7), 2755 (2020)
12. Wang, Y., Xie, T., Jie, X.: A mathematical analysis for the forecast research on tourism carrying capacity to promote the effective and sustainable development of tourism. Discret. Contin. Dynam. Syst. **12**, 837–847 (2018)
13. Muneeb, S.M., Nomani, M.A., Asim, Z., Adhami, A.: Assessing and optimizing decision-making policies of India with public employment growth as a key indicator toward sustainable development goals using multicriteria mathematical modeling. J. Public Aff. **22** (2021)
14. Lafuente-Lechuga, M., Cifuentes-Faura, J., Faura-Martínez, Ú.: Mathematics applied to the economy and sustainable development goals: a necessary relationship of dependence. Educ. Sci. **10**(11), 339 (2020)
15. Celli, I., et al.: Development of a tool to optimize economic and environmental feasibility of food waste chains. Biomass Convers. Biorefinery **12**(10), 4307–4320 (2022)
16. Egelston, A., Cook, S., Nguyen, T., Shaffer, S.: Networks for the future: a mathematical network analysis of the partnership data for sustainable development goals. Sustainability **11**(19), 5511 (2019)
17. Liu, W., Hou, Q., Xie, Z., Mai, X.: Urban network and regions in China: an analysis of daily migration with complex networks model. Sustainability (Switzerland) **12**(8), 3208 (2020)
18. Jyotsna, K., Tandon, A.: A nonlinear mathematical model investigating the sustainability of an urban system in the presence of haphazard urban development and excessive pollution. Nat. Resour. Model. **31**(2), e12163 (2018)
19. Suvorov, N., Suvorova, I.: Discovering the mathematical formula of the universal law of harmony of the CREATOR and the law of harmony of GOD to ensure a holistic sustainable development of man and humanity. Eur. J. Sustain. Dev. **7**(2), 81 (2018)
20. Mazur, B., Walczyna, A.: Bridging sustainable human resource management and corporate sustainability. Sustainability (Switzerland) **12**(21), 8987 (2020)
21. Cristina, I.O.M., Nicoleta, C., Cătălin, D.R., Margareta, F.: Regional development in Romania: empirical evidence regarding the factors for measuring a prosperous and sustainable economy. Sustainability (Switzerland) **13**(7), 3942 (2021)
22. Melkonyan, A., et al.: Integrated urban mobility policies in metropolitan areas: a system dynamics approach for the Rhine-Ruhr metropolitan region in Germany. Sustain. Cities Soc. **61**, 102358 (2020)

23. Scheel, C., Aguiñaga, E., Bello, B.: Decoupling economic development from the consumption of finite resources using circular economy. A model for developing countries. Sustainability (Switzerland) **12**(4), 1291 (2020)
24. Wu, H., Fan, W., Lu, J.: Researching on the sustainability of transportation industry based on a coupled emergy and system dynamics model: a case study of Qinghai. Sustainability (Switzerland) **13**(12), 6804 (2021)
25. Rabe, M., Streimikiene, B., Bilan, Y.: The concept of risk and possibilities of application of mathematical methods in supporting decision making for sustainable energy development. Sustainability **11**(4), 1–24 (2019)
26. Chen, A., et al.: Spatiotemporal dynamics of ecosystem fires and biomass burning-induced carbon emissions in China over the past two decades. Geogr. Sustain. **1**(1), 47–58 (2020)
27. Krapivin, V.F., Mkrtchyan, F.A., Rochon, G.L.: Hydrological model for sustainable development in the Aral Sea Region. Hydrology **6**(4), 91 (2019)
28. Fleacă, E., Fleacă, B., Maiduc, S.: Aligning strategy with sustainable development goals (SDGs): process scoping diagram for entrepreneurial higher education institutions (HEIs). Sustainability (Switzerland) **10**(4), 1032 (2018)
29. Website of Federal state statistics service. https://rosstat.gov.ru/. Accessed 01 Feb 2023

Solving the Issue of Managerial Decision Making in Terms of Ranking by the Effective Rank Method

Natalia E. Buletova[1(✉)], Ekaterina V. Stepanova[2], Natalia N. Kulikova[3], and Galina V. Kuzibetskaya[4]

[1] Russian Academy of National Economy and Public Administration (RANEPA), 82, Vernadsky Avenue, Moscow, Russia
buletovanata@gmail.com

[2] Volgograd Institute of Management, Branch of RANEPA, 8, Gagarin Street, Volgograd, Russia

[3] Russian Technological University (RTU MIREA), 78, Vernadsky Avenue, Moscow, Russia

[4] Plekhanov Russian University of Economics, Volgograd Branch, 11, Volgodonskaya Street, Volgograd, Russia

Abstract. Managerial decision-making process is dependent on complete or asymmetric information that determines logic, reliability and results of the corporate management. Business administration requires the most accurate application of information and communication technologies with possibility of automated decision-making included in the methodology for calculating the rating of the managed items or objects. In the paper, the authors-developed Method of rank analysis was developed and applied based on the replacement of ordinal ranking with the effective series in the set of units or objects of management belonging to a homogeneous set (enterprises and organizations of a particular industry, the territory of a particular country, employees of an enterprise, etc.). This study is based on the following assumptions: whatever rating technique is applied in the analysis and management of this homogeneous set, as a result, a Top 10 or other similar list is obtained that distributes the rated units in an ordinal series. Secondly, incorrectness of such a distribution is obvious and requires a change in the ordinal ranking, namely, its replacement with an effective ranking. By doing so, the latter takes into account the degree of difference in the rating values of the selected objects of analysis and management. Thus, the authors propose to use two approaches to constructing an effective rank: 1) based upon the use of the normalized value formula of the so-called "Linear Normalization Model" (closed ranking scale) and 2) in accordance with linear rank distribution of objects modeling to which one or more rating is applied (e.g. rating of investment attractiveness, quality of life, digitalization, etc.) which characterizes the main (central) part of the objects producing the largest R^2 values processing their own calculation system (open ranking scale). As a result, experts responsible for monitoring the system and the managerial decision-making algorithm will obtain the most reliable information on the actual distribution of the ranking objects, taking into account the observed differences in the received ratings of the adjacent rating objects.

Keywords: Rating · Ordinal Rank · Effective rank · Rank Distribution · Closed Ranking Scale · Open Ranking Scale

© The Author(s), under exclusive license to Springer Nature Switzerland AG 2023
A. Gibadullin (Ed.): DITEM 2022, LNNS 683, pp. 90–100, 2023.
https://doi.org/10.1007/978-3-031-30926-7_9

1 Introduction

Prevalence of ratings as a way to present expert assessments results of managed objects, planning and investment allows us to talk about the presence of a serious error in the application of any rating. The error occurs in the ordinal ranking of the rated objects, when despite different rating values of the adjacent objects, they receive ordinal ranks. Therefore, all the participants of the rating are misinformed, including, firstly, the customer who makes important administrative decisions based on the obtained ranking data, secondly, the representative of the rating object who does not get an appropriate assessment of its position in the competitive market or real differences in the ranked indicators with adjacent objects. Such differences are mostly evident on the examples of global economies ranked by GDP or GDP based on purchasing power parity, when countries included in the Top 5 or Top 10 largest economies within the ratings have one or more levels of gradation among themselves according to the degree of difference (possibly many times greater) in the values of the ranked quantities.

In modern practice, questions of the ordinal ranking effectiveness and of correctness of the data for monitoring, evaluating and making managerial and investment decisions are not raised due to the availability of detailed information for each object included in the rating. Nevertheless, clarity and visibility in the application of an effective rank, which reflects the degree of difference between adjacent objects in the rating, allows not only to improve the quality of ranking as such, but also to provide the management team with a reliable monitoring tool for any asset managed in order to make motivated and accurate decisions.

The effective rank method can be applied in the course of a company's self-assessment within the framework of the European Foundation for Quality Management model (EFQM-2020) that encompasses sustainability and shares features with Management 3.0. The method measures an effective dimensionality when summarizing the evaluation results for all the parties concerned including staff, management and society. Additionally, the effective rank method can be used for competitions implemented within the framework of the Business excellence model, which in addition to European and other similar contests also include the Competition for awards of the government of the Russian Federation in the field of quality.

2 Materials and Methods

The knowledge base for the study is the open data of various ratings used by experts to assess and monitor management objects (e.g. top managers rating of enterprises in a particular industry). Furthermore, in order to test and compare the results of the Effective rank method application, the authors used materials from various official sources which use different assessment techniques, for example, a performance metrics of social and economic development of Russian constituent entities or an indicator framework calculating global economies' GDP per capita data.

The authors of the study reviewed and took into account research materials of Russian and international scientific community who, in their works, had observed the issue of rating application in evaluation, monitoring and making managerial decisions when considering investment objectives, increased control, innovations or support measures and had presented the developed and applied rating results in management. This was done for the purpose of substantiating the relevance of the research topic and proving its prevalence in the scientific community. First of all, the studies by Brueckner [3], previous research dating back to the 1990s by Bergeron, Morris, Banegas [2], Deadrick, Gardner [5], Schwarzwald, Koslowsky and Mager-Bibi [10] should be noted as they considered historical evolution of scientific thought on ratings and other ways of evaluation and comparison of certain objects of research. More recent studies by Holla, Kavitha [6], Khalafyan, Shevchenko [7], Lavrova [8], Rafida, Widiyatni, Harpad, Yulsilviana [9], Yakimets, Kurochkin [11], Zaznobina, Molkova, Basurov, Gelashvili [12] and Zhgun [13] corroborate the ideas on evaluation importance and the rating scale used in various areas of professional, economic activity in assessing the development results of territories, regions, integration groupings of countries for the allocation of units in the aggregate based on the estimates. However, traditional ordinal distribution of leaders and all other rating objects is misleading. For instance, distorted estimated indicators are incorrectly applied in assessing competition results including the description of the EFQM 2020 Model in [1]).

To the traditional methods of economic analysis (modeling, comparison, analogy, factor analysis), the authors of the study have added description and the effective rank methods; a part of the effective rank method was patented in 2017 and has proved its reliability and correctness in application and testing (an example of the works by N. Buletova, E. Stepanova [4]). Traditional methods of economic and mathematical modeling were of great importance for the study, including normalizing actual values, that is bringing the initial data to acceptable intervals of numerical values.

3 Results

3.1 Modeling the Process of Ranking Distribution of Management Objects

In compliance with the research goal and expected results of enhancing the approach to rating objects distribution by effective ranks, we will present the main characteristics of the Rank distribution model for the management objects with particular values of the ranked attribute ranging from x_{min} to x_{max}.

It is important to clarify that in order to identify linear and non-linear dependencies observed in the ranked values, it is necessary to use a sufficient number of rating objects (regions, enterprises, industry leaders, employees of the same specialty, etc.).

Another important reasoning for the applied Model is that among these rating objects, the participant with the highest (best) value of the ranked attribute will rank 1 and the participant with the lowest (worst) value of the ranked attribute will have the last maximum rank R, respectively.

As a result, we present the function $x = f(R)$ which reflects the essence of the ranking distribution of rating objects by value of feature x.

It has been empirically determined that there are two possible approaches to measuring and presenting the effective rank values, which reflect the difference degree among adjacent rating objects.

The first approach is founded on the application of a "Linear normalization model" which in, its turn, uses a normalization formula (Max-Min Normalization, Min-Max Scaling) based on the transformation of feature values to change original feature values by transforming them in the range of values from 0 to 1:

$$\tilde{x} = \frac{x_i - x_{\min}}{x_{\max} - x_{\min}} \tag{1}$$

It is necessary to ensure graduation of the extreme deviations of the ranked features, and the rank as an integer will be defined as *f(x)*. As a result, experts get the opportunity not only to number the rating objects from *1* to the *n* in accordance with their number, but also to take into account their positional deviation considering the degree of differences in the achieved analyzed development indicator. The authors' addition to this formula will be the use of not the initial values $\mathrm{x_{min}}$, $\mathrm{x_{max}}$, but of those in trend:

$$\tilde{x} = \frac{x_i - x_{\min(trend)}}{x_{\max(trend)} - x_{\min(trend)}} \tag{2}$$

where $x_{min\ (in\ trend)}$ and $x_{min\ (in\ trend)}$) are the minimum and maximum normalized values found using the linear trend formula (see Fig. 1).

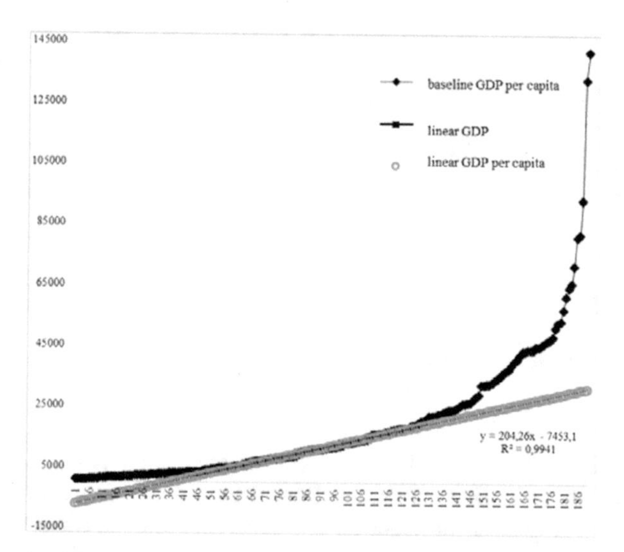

Fig. 1. Example of Empirical rank distribution of the global economies by GDP per capita.

In order to convert the ranking values into integers and to build the true distribution of regions, in this study it is proposed to take into account the degree of discrepancy among the values $\tilde{x}_i$, and then the ranks that will be assigned to the rating objects according to their place in the range of normalized values will have a real deviation (measure of spread) among themselves, showing the actual position of the territorial entity of the RF:

- Minimum rank is 0; maximum rank is, for example, 200;
- Normalized values $\tilde{x}_i$ are converted to integers (Table 1).

Table 1. Example of Ranking objects with deviation of integer ranks on scale from 0 to 200.

Ranking object									
Normalized values	0.000	0.070	0.153	0.167		2.392	2.614	3.115	3.170
The ratio of x_i to x_{max}	0	0.022	0.048	0.053	……	0.754	0.825	0.982	1
RANK with deviation on a scale from 0 to 200	0	4	10	11		151	165	196	200

Figure 2 visualizes distribution of the ranks of the objects presented in Table 1 with a deviation demonstrating differences in the achieved development results.

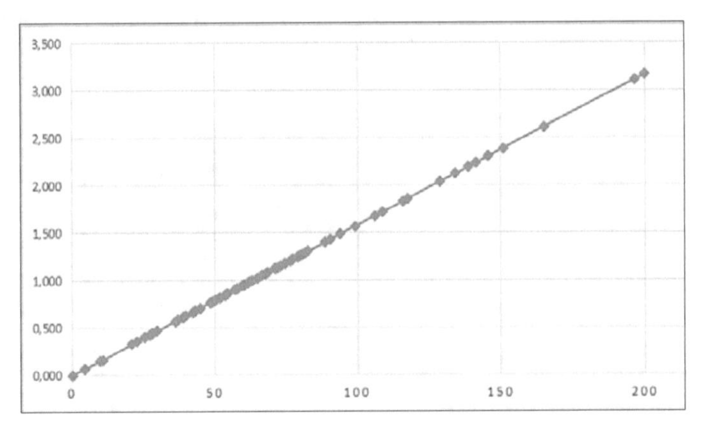

Fig. 2. Example of ranking distribution of rating objects on a Rank scale from 0 to 200 (Closed ranking scale).

However, while maintaining the limit value of ranks in the amount of 200, the presented approach does not allow estimating the changing slope of the curve with a varying degree of non-linearity in the distribution of objects according to a chosen development indicator.

3.2 Linear Rank Distribution

Since an effective rank construction in the Linear normalization model is possible in the format of a closed scale (x_{max} value as the formula parameter can be determined by experts), to continue modeling the rank distribution, we will present an alternative option with an open ranking scale (maximum rank value does not depend on the number of rating objects or results of the normalization of the ranked values. It is calculated in the following way:

- The variation range is determined according to the actual ranked values that the rating objects possess:

$$R = x_{\max} - x_{\min} \tag{3}$$

- A linear rank distribution is modeled:

$$LRD = x_{\min} + (i - 1) \cdot h \tag{4}$$

  where $i = 1,\ldots,$ n is the rank r of the rating object (staff of a company, organizations of one field of activity, world countries);
  $h = \mathrm{R}/(\mathrm{n} - 1)$.
- An isomorphic mapping of the non-linear structure of the values of the indicator selected for ranking is built into the corresponding structure of positive integers:

$$y = k \times r + a_0 \tag{5}$$

  where y is the indicator used as the basis for the object rating (ranked value);
  k is a linear coefficient;
  r is the region rank;
  a_0 is an intercept term.
- Inserting empirical values of the ranked indicator in the previous equation instead of y, and resolving it with respect to the rank r, we obtain the following expression for rank determining:

$$r = Integer\left[\frac{y - a_0}{k}\right] \tag{6}$$

  where *Integer (argument)* is a function that rounds its *argument* to the nearest smaller integer value.
- The formula representing isomorphic mapping in item 3 is transformed into a formula for calculating the effective rank r*, where instead of k and a_0, the parameters of the straight line equation calculated from the linear segment are substituted:

$$y = k \times r* + a_0 \tag{7}$$

- In order to perceive correctly values of effective ranks, it is recommended to convert them taking into account traditional representation of Top-5, Top-10, where the best rating value is assigned an ordinal rank of 1, therefore, an additional procedure for building effective rank values for all rating objects is the need to shift the minimum rank to value of 1.

Considering the application scope of the presented methodology, we can make a conclusion on the relevance of the results obtained which differ from the current ranking practice in greater accuracy and reliability. The results obtained are relevant not only for effective presentation of management results of objects monitoring, but also for reflecting the outcomes of the organization's self-assessment for ranking participants in the course of development or transformation.

This method of calculating the effective rank may be of particular interest for:

- Investors of any scale or capital financing areas in order to prevent the use of asymmetric information that distorts the perception of the investment object and evaluation of capital investment efficiency;
- Company strategic development requiring making necessary critical changes based on the ranking results, estimating corporate personnel performance, top management efficiency, achieving company's "soft" goals including those related to multi-level leadership and its identification by ranks for participants.
- Government agencies responsible for the development and application of national/regional ratings, which are being introduced into the system for assessing professional activity, investment territory attractiveness, etc.

Table 2 provides an example of a simulated application of the effective rank method using a linear rank distribution.

Table 2. Modeling the construction of ordinal and effective ranks system reflecting the result of rating objects (*from 1 to N, N = 15*).

Rating objects	Ranked indicator value (result of rating method application)	Ordinal rank assigned to the rating objects, r	Effective rank, r^*	
			Reference period	Reporting period
1	81.19	1	1	1
2	52.58	2	11	13
3	50.37	3	12	15
...	...	...	...	...
N	17.35	15	24	27

- In the modeled example, there are *N* (15) units of managed objects that were included in the rating and distributed according to the obtained rating values ($x_{max} = 81.19$; $x_{min} = 17.35$);
- When applying the distribution of participants' rating according to the traditional ordinal distribution, the number of ranks r coincides with the number of rating objects (N), while in the example, to demonstrate the possibilities of the effective rank method, the values of the effective rank r^* are presented in dynamics for reference and reporting periods;
- Comparing the values of the ordinal r and effective r^* ranks, it can be seen that the gap between the objects that took 1st and 2nd positions in the rating is not one ordinal number, as is perceived in the traditional distribution of participants in any rating, but differs by 10 units (effective ranks 1 and 11 for the first and second objects of rating according to the example of the reference period);

- Open ranking scale (see Fig. 3) is an important feature of the linear rank distribution model: the maximum rank value will change depending on differences varying degree among the ranked values (following the change in the values of the x_{max} ranked indicator).

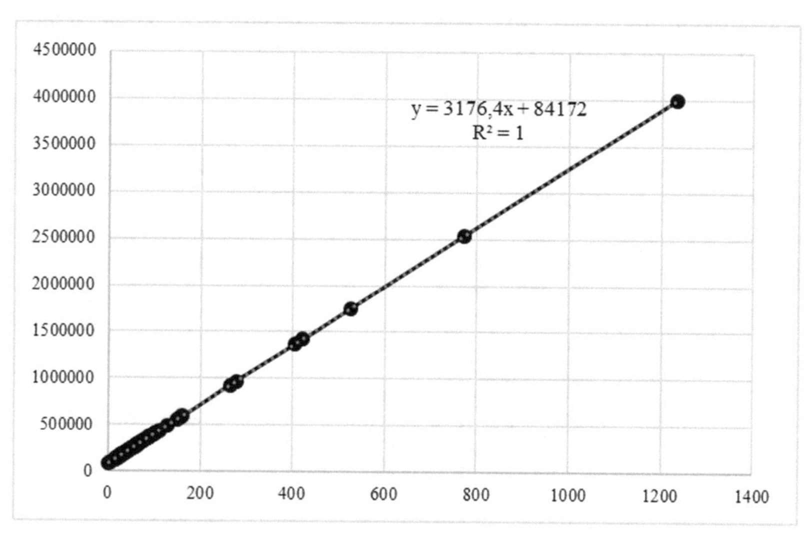

Fig. 3. Example of the Ranking distribution of rating objects using a linear isomorphic mapping $y = k * y + a_0$ (open ranking scale).

4 Discussion

The effective rank method application cannot be fully automated, since one of the stages of applying the linear rank distribution requires the participation of an expert in determining the so-called "linear section" in the observed set of values of the ranked indicator for the rating objects set. The selection is carried out in order to determine the population for which the determination coefficient R^2 value will be as close as possible to 1. However, using a computer program "Effective Rank Calculation of Objects by their Parameter Values", certificate of state registration No. 2017619388 obtained by the author in 2017, the biggest part of the Method can be automated.

The experience of applying the Method has provided relevant finding and proved:

- Its application scalability for different rating objects expressed, among other things, in the possibility of compiling a new rating of objects by accumulating the n-th number of ratings of one object through the sum of effective ranks related to one period;
- Greater efficiency of the Linear rank distribution Model for an open scale when determining effective ranks, because, when comparing calculation results of different periods, changes in the rank on an open scale and an assessment of the magnitude of the

so-called shift in the value of the effective rank of a single object rating over time are important factors determining the objects' development outcomes;
- Provided that for one set of objects of management several ratings or indicators are used for ranking, it is possible to build a new rating based on a simple summation of 2 or more effective ranks r_i*, this will allow not only to streamline the use of several values of effective ranks obtained in the analysis process objects, but also to get a new monitoring and analysis tool based on the aggregation of different ratings of the managed population (Table 3).

Table 3. Example of rating formation of management objects by Method of aggregating two or more effective ranks r_i* according to the ranked values system.

Objects of Management	Effective ranks built on values of m-th number of ratings applied to management objects			Rating of Objects of Management	
	r_1*	r_2*	r_m*	sum of ranks r_i*	shift - ranking from 1 to r_{max}
1	1	21	58	80	1*
2	462	206	95	763	684
3	831	22	78	931	852
...	...	...			
N	1204	11	119	1334	1255

* it is necessary to "shift" all rating values from the minimum (for example 80) to 1, that is, take a step in shifting the rating by 79 units.

The obtained rating makes it possible to receive the following results and interpretations of management objects performance (both individually and for the entire population):

- Ranks analysis obtained for each indicator is primary, since only in this case it is possible to obtain a detailed interpretation of each object including in comparison with others from the same set;
- New rating of management objects, defined as the sum of effective ranks, allows to provide an independent rating assessment of any management object regardless of the goals and authorship of the medical rating;
- Effective ranks summation provides a cumulative effect in determining the results of the management objects assessment and of the entire estimated population, and most importantly, scalability and universality of comparing the obtained rating values in time and space.

5 Conclusion

Reliability increase of evaluation features focused on adjusting the results of rating application (for the purposes of conduction quality and performance competitions among organizations within the EFQM-2020 Model or internal assessment of management quality, personnel rating for incentives, etc.) requires improving the approach to winners' distribution. Adjustments are necessary as even among the Top 5 or Top 10 rating objects there are significant differences in the achieved value of the compared indicator/indicators.

These results provide further support for the hypothesis that application of the Effective rank method presented in the study for new rating compilation based on the obtained set of effective ranks is of high efficiency, since it characterizes the rating results according to two or more criteria and accumulates all ratings results in one set of objects in general conditions of place and time. All these factors increase reliability of the rank information used in managerial decision-making, reduce the data asymmetry and allow improving the efficiency of both assessment procedures and outcomes of the entire cycle of the objects' performance, managing system and monitoring.

References

1. Akatov, N.B., Safonov, A.A., Bryukhanov, D.: EFQM 2020 model: focus on outstanding results. Qual. Manag. Methods **11**, 10–15 (2019)
2. Bergeron, G., Morris, S.S., Banegas, J.M.M.: How reliable are group informant ratings? A test of food security ratings in Honduras. World Dev. **10**, 1893–1902 (1998)
3. Brueckner, L.J.: Chapter XII: rating scales, score-cards, and checklists. Rev. Educ. Res. **9**(5), 524–527 (1939)
4. Buletova, N.E., Stepanova, E.V.: Structural differences of economies of different developmental types: assessments and effective ranking on a global basis. Int. J. Appl. Eng. **12**(22), 12554–12563 (2017)
5. Deadrick, D.L., Gardner, D.G.: Distributional ratings of performance levels and variability: an examination on rating validity in a field setting. Group Organ. Manag. **23**, 317–342 (1997)
6. Holla, L., Kavitha, K.S.: A sentiment score and a rating based numeric analysis recommendations system: a review. Recent Adv. Comput. Sci. Commun. **14**(1), 236–245 (2021)
7. Khalafyan, A.A., Shevchenko, I.V.: Compilation and assessment of the consistency of bank ratings by means of computer analysis. Finance Credit **28**(748), 1655–1677 (2017)
8. Lavrova, A.P.: Rating assessment of the region's competitiveness. Proc. Int. Acad. Agrarian Educ. **59**, 105–110 (2022)
9. Rafida, V., Widiyatni, W., Harpad, B., Yulsilviana, E.: Implementation of multi-attribute rating technique in selection of acceptance scholarship of PMDK (case study: stmik widya cipta dharma). Int. J. Mod. Educ. Comput. Sci. **13**(1), 22–33 (2021)
10. Schwarzwald, J., Koslowsky, M., Mager-Bibi, T.: Peer ratings versus peer nominations during training as predictors of actual performance criteria. J. Appl. Behav. Sci. **3**, 360–372 (1999)
11. Yakimets, V.N., Kurochkin, I.I.: Analysis of results of the rating of volunteer distributed computing projects. Commun. Comput. Inf. Sci. **965**, 472–486 (2019)

12. Zaznobina, N.I., Molkova, E.D., Basurov, V.A., Gelashvili, D.B.: Rating analysis of the BRICS countries in terms of socio-ecological and economic indicators based on the generalized desirability function. Probl. Reg. Ecol. **3**, 137–142 (2018)
13. Zhgun, T.V.: Method for evaluating the robustness of rankings generated by composite indices. J. Phys. Conf. Ser. **1352**(1), 012065 (2019)

Methodology for Evaluating the Economic Efficiency of Using VR-Projects in the Activities of Industrial Companies

Ekaterina N. Kharitonova[1(✉)], Natalia N. Kharitonova[1], Ilia A. Litvinov[2], and Sadriddin R. Chorshanbiev[3]

[1] Financial University under the Government of the Russian Federation, 49, Leningradsky Pr., Moscow 125167, Russia
eharitonova@fa.ru

[2] Limited Liability Company "Interkos-IV" (PJSC Magnitogorsk Iron & Steel Works and Subsidiaries), Office 334, House 122, Lit. A, Territory of Izhora Plant, Kolpino, St. Petersburg 196650, Russia

[3] Tajik National University, 17, Rudaki Avenue, Dushanbe 734025, Republic of Tajikistan

Abstract. In modern economic conditions, virtual reality is beginning to be used not only in the gaming industry, but also in the business processes of industrial companies. The article provides a brief overview of the forecast data for the virtual reality market until 2028, discusses the key players in this market (hardware and software manufacturers, as well as manufacturers of VR headsets). In 2022, the Financial University (Moscow) carried out a scientific study on the topic "The use of virtual reality elements in the company's environmental business processes" (registration No. NIOKTR 122012100228-9), which studied the activities of 156 largest industrial companies in Russia from 17 different industries, whose business processes began to use various VR projects. In general, out of 156 industrial companies studied, 58 companies use various virtual reality projects (from a training program to a "digital factory") in their activities, including 3 non-ferrous metallurgy companies and 6 ferrous metallurgy companies. The authors propose to use several models to assess the economic efficiency of VR projects of industrial companies that allow calculating profits and costs both for individual VR projects and in general for all VR projects used in the business processes of the companies under study.

Keywords: Activities of Industrial Companies · Methodology for Evaluating Economic Efficiency · VR-projects

1 Introduction

The third decade of the 21st century is a time of new economic and political challenges, new technologies and new business opportunities within the framework of modern strategies. One of the relatively "new" technologies can be considered virtual reality (abbreviated as VR). "VR is a digitally created experience where a three-dimensional

© The Author(s), under exclusive license to Springer Nature Switzerland AG 2023
A. Gibadullin (Ed.): DITEM 2022, LNNS 683, pp. 101–111, 2023.
https://doi.org/10.1007/978-3-031-30926-7_10

environment is simulated with the real-world. The technology offers an immersive experience to the viewers with the help of VR devices, such as headsets or glasses, gloves, and bodysuits. The technology has brought a transformation in the gaming and entertainment industries by allowing users to experience immersion in a highly virtual realm" [1].

Virtual reality is a modern innovative technology that allows you to "create a new world". Its use is not limited to the gaming industry. VR projects are actively developing in various industries and services. "In addition, the increasing usage of this technology in instructive training, such as for training mechanics, engineers, pilots, soldiers in defense, field workers, and technicians, in the oil & gas and manufacturing sectors is driving the market growth" [1].

"The global VR market is expected to grow at a compound annual growth rate of 18.0% from 2021 to 2028 to reach USD 69.60 billion by 2028" (see Fig. 1) [1].

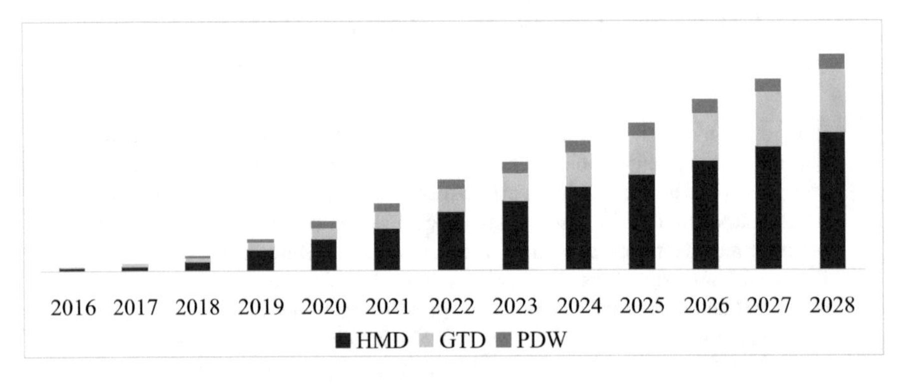

Fig. 1. The size of the virtual reality market (by device), in 2016–2028, million US dollars [1].

"The VR technology is beneficial for product prototyping or training with the help of its immersive 3D technology. The segment can be categorized into software types, such as simulation, training, virtual tool, game development, application, a learning experience platform, gamification, and segment reality" [11].

In 2022, the authors conducted a scientific study[1] related to the study of the experience of Russian industrial companies in the field of VR projects [2].

The largest participants in the VR market (manufacturers of hardware and software products) are shown in Tables 1 and 2.

The global virtual reality headset market size was valued at USD 7.81 billion in 2020 and is expected to grow at a compound annual growth rate (CAGR) of 28.2% from 2021 to 2028 (see Fig. 2) [17].

"In addition, the incorporation of Artificial Intelligence (AI) and machine learning into VR is likely to fuel the product demand. The key players in the industry are continually investing in R&D for the development of state-of-the-art devices" [17].

[1] At the Financial University in 2022, a research work (hereinafter referred to as R&D) was carried out on the topic: "The use of virtual reality elements in the company's environmental business processes". Registration number NIOKTR 122012100228-9.

Table 1. World leaders in the VR market - equipment manufacturers.

No.	Names of the companies	Their official websites on the Internet
1	Kopin Corporation	https://www.kopin.com/about/ [3, p. 1]
2	Lenovo Group Limited	https://www.lenovo.com/ru/ru/about/ [4, p. 1]
3	Samsung Electronics Co. Ltd.	https://www.samsung.com/ru/ [5, p. 1]
4	Epson Corporation	https://global.epson.com/ [6, p. 1]
5	Sony Corporation	https://www.sony.com/en [7, p. 1]
6	Vuzix	https://www.vuzix.com/ [8, p. 1]

Table 2. World leaders in the VR market - manufacturers of software products.

No.	Names of the companies	Their official websites on the Internet
1	Apple	https://www.apple.com/ru/ [9, p. 1]
2	Blippar	https://www.blippar.com/ [10, p. 1]
3	Google	https://about.google/ [11, p. 1]
4	Magic Leap	https://www.magicleap.com/en-us [12, p. 1]
5	Microsoft Corporation	https://www.microsoft.com/ru-ru [13, p. 1]
6	Niantic Inc.	https://nianticlabs.com/ [14, p. 1]
7	Sony Corporation	https://www.sony.com/en [7, p. 1]
8	Wikitude	https://www.wikitude.com/ [15, p. 1]
9	Zappar	https://www.zappar.com/ [16, p. 1]

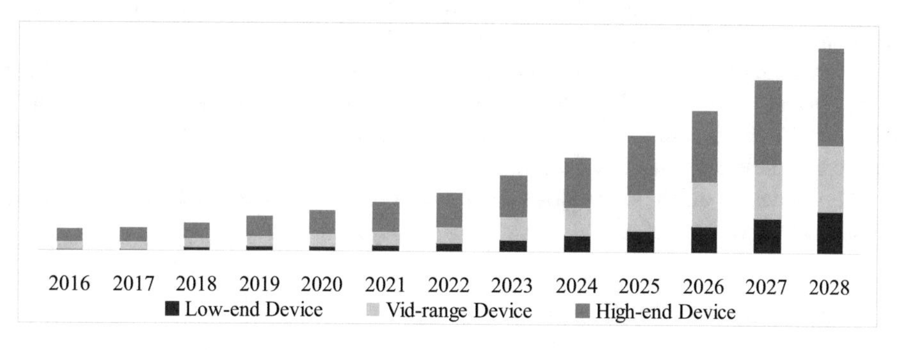

Fig. 2. VR headset market size by end device, 2016–2028, million US dollars [17].

Some notable players in the global VR headset market include the following companies (See Table 3).

The VR market in Russia is also developing quite actively. Decree of the President of the Russian Federation of May 9, 2017 No. 203 "On the Strategy for the Development of the Information Society in the Russian Federation for 2017–2030" [23] defines the development of information and communication technologies, including VR technologies.

At the state level, as a whole, the "Roadmap for the development of "end-to-end" digital technology "Virtual and Augmented Reality Technologies" has been approved and is being actively implemented [24]. The implementation of various VR and AR projects is envisaged until the end of 2024.

Table 3. World market leaders in VR headsets.

No.	Names of the companies	Their official websites on the Internet
1	Carl Zeiss AG	https://www.zeiss.ru/corporate/ru/home.html [18, p. 1]
2	Facebook Technologies, LLC (Oculus)	https://www.oculus.com/ [19, p. 1]
3	Google LLC	https://about.google/ [11]
4	HTC Corporation	https://www.htc.com/eu/ [20, p. 1]
5	LG Electronics	https://www.lg.com/ru [21, p. 1]
6	Microsoft Corporation	https://www.microsoft.com/ru-ru [13, p. 1]
7	Razer Inc.	https://www.razer.ru/ [22, p. 1]
8	Samsung Electronics Co., Ltd.	https://www.samsung.com/ru/ [5, p. 1]
9	Sony Corporation	https://www.sony.com/en [7, p. 1]

The authors studied the activities of 156 largest Russian industrial companies from the Expert-400 rating, belonging to 17 industrial sectors [25] (see Table 4).

More than half of companies from many industrial sectors ("Forestry, woodworking and pulp and paper industries", "Light industry", "Manufacture of weapons and ammunition", "Mechanical engineering", "Printing industry") are actively implementing VR projects in their activities (see Fig. 3).

The structure of 58 VR projects of the largest industrial companies in Russia by industry is shown in Fig. 4.

2 Materials and Methods

Let us consider in more detail the VR projects implemented by Russian metallurgical companies (see Tables 5 and 6).

Table 4. Number of VR-projects for 156 largest Russian industrial companies included in the Expert-400 rating.

No.	Sector of industry[2]	Number of companies	VR-projects	Share of VR-projects in the total number of companies
1	Building materials industry	4	1	25.0%
2	Chemical and petrochemical industry	16	6	37.5%
3	Coal industry	6	2	33.3%
4	Diversified holdings	2	2	100.0%
5	Ferrous metallurgy	14	6	42.9%
6	Food industry	19	3	15.8%
7	Forestry, woodworking and pulp and paper industries	5	3	60.0%
8	Industry of precious metals and diamonds	5	1	20.0%
9	Light industry	1	1	100.0%
10	Manufacture of weapons and ammunition	3	2	66.7%
11	Mechanical engineering	30	17	56.7%
12	Non-ferrous metallurgy	7	3	42.9%
13	Oil and gas industry	19	6	31.6%
14	Pharmaceutical industry	1	0	0.0%
15	Power engineering	18	4	22.2%
16	Printing industry	1	1	100.0%
17	Tobacco industry	5	0	0.0%
	Total	156	58	37.2%

The authors propose to use the following model (1) to assess the economic efficiency of VR projects of an industrial company:

$$EE_{VR-project_i} = \frac{P_{VR-project_i}}{C_{VR-project_i}} * 100\% \quad (1)$$

[2] The list of industries in Table 1 is presented in alphabetical order.

Table 5. Virtual reality projects implemented by major Russian non-ferrous metallurgy companies.

No.	No. From the Expert-400 rating	Company name, official website	Name of the VR-project	Essence of the VR-project
1	11	"Norilsk Nickel", https://www.nornickel.ru [26]	"Virtual factory"	The Sulfur Project Production Line of the Copper Plant in VR: Tutorial
2	21	"Rusal", http://rusal.ru [27]	Casting Machine Virtual Simulator	Interactive computer system for personnel training. The functionality of this system is implemented on the basis of realistic 3D models of the casting unit using VR technologies
3	22	"UGMC Group", https://www.ugmk.com [28]	VR exposure	Neural networks that, after a certain training, are able to process Big Data with a given result clearly and without errors

where $EE_{VR-project_i}$ (economic efficiency of VR-project) – economic efficiency of a specific (i-th) VR project of an industrial company,

$P_{VR-project_i}$ (profit from VR-project) – profit of an industrial company received from the implementation of a specific (i-th) VR project,

$C_{VR-project_i}$ (VR-project costs) – costs of an industrial company for the implementation of a specific (i-th) VR project.

At the same time, to determine the total amount of profit (economic benefits) from the implementation of all VR projects of an industrial company and the total amount of costs for the implementation of these VR projects, it is advisable to use the following additive models (2) and (3):

$$P_{VR-projects} = \sum_{i=1}^{n} P_{VR-project_i} \tag{2}$$

$$C_{VR-projects} = \sum_{i=1}^{n} C_{VR-project_i} \tag{3}$$

where $P_{VR-projects}$ (*profit from VR-projects*) – profit of an industrial company received from the implementation of all its VR projects,

$P_{VR-project_i}$ (*profit from VR-project*) – profit of an industrial company received from the implementation of a specific (i-th) VR project,

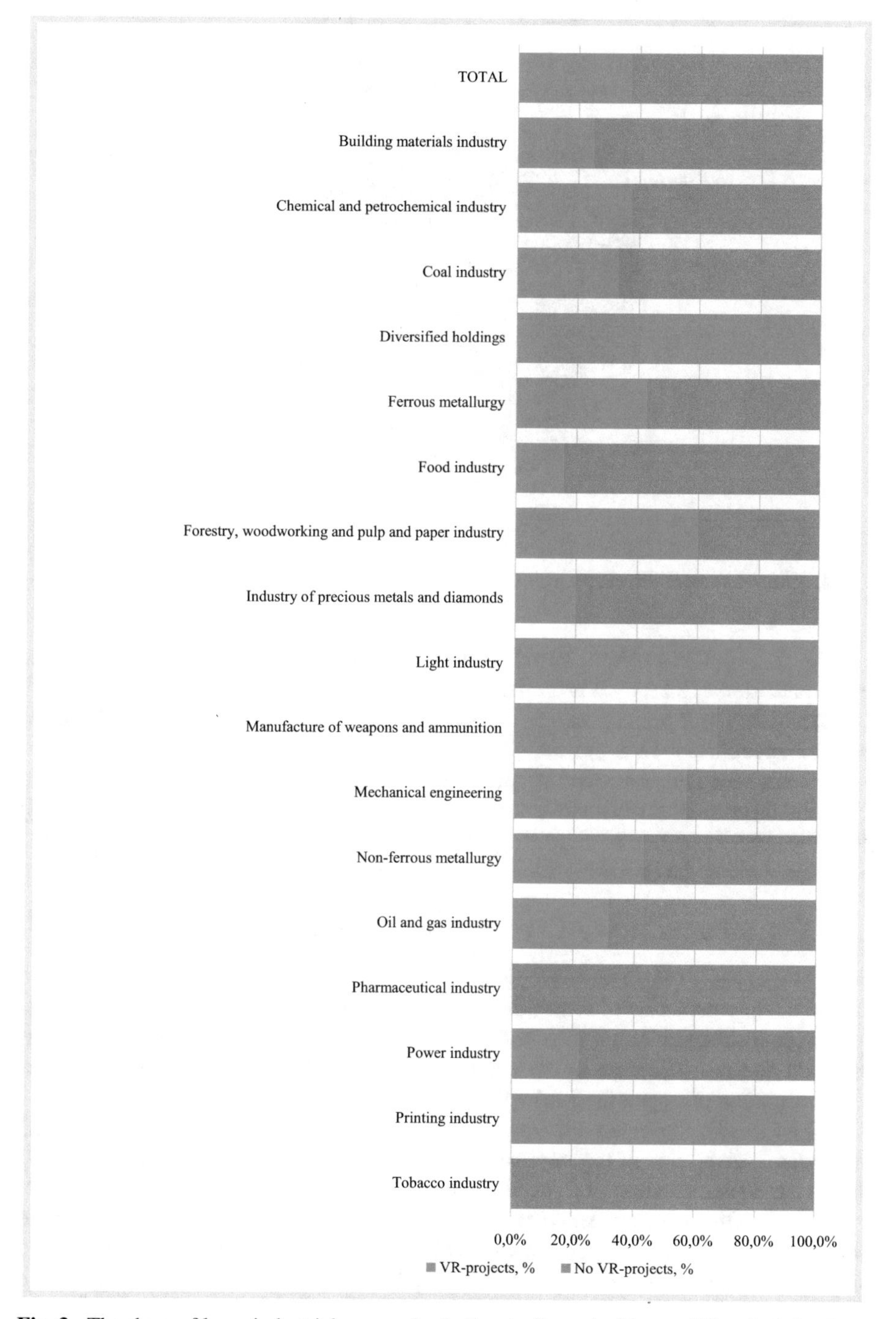

Fig. 3. The share of large industrial companies in Russia (from the Expert-400 rating) that have implemented VR projects in their business processes.

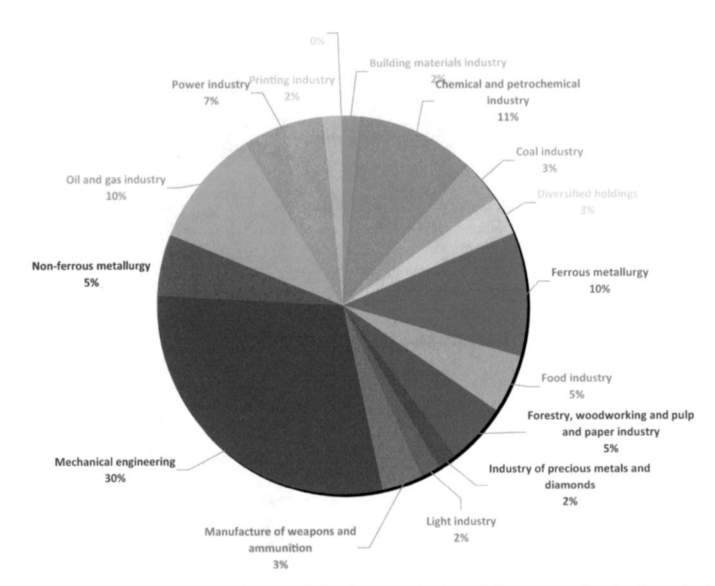

Fig. 4. The structure of 58 VR projects of the largest industrial companies in Russia by industry.

$C_{VR-projects}$ (*VR-projects costs*) – the costs of an industrial company for the implementation of all its VR projects,

$C_{VR-project_i}$ (*VR-project costs*) – costs of an industrial company for the implementation of a specific (i-th) VR project,

n – number of implemented VR projects of an industrial company.

It should be emphasized that the total amount of profit (or economic benefits received) from the implementation of a specific (i-th) VR project of an industrial company can be determined using the following additive model (4):

$$P_{VR-project_i} = P_{RM\&I} + P_{ICPP} + P_{CS\&ATR} + P_{DNR} + P_{ILP} + P_{RC} + P_{\mathrm{MM}} \quad (4)$$

where $P_{VR-project_i}$ (*profit from VR-project*) – profit of an industrial company received from the implementation of a specific (i-th) VR project,

$P_{RM\&I}$ (profit, reduce marriage and inconsistencies) – profit of an industrial company, obtained due to a decrease in the number of defects and inconsistencies due to the implementation of a specific (i-th) VR project,

P_{ICPP} (*profit, improve consumer properties of products*) – profit of an industrial company received due to the improvement of the consumer properties of its products due to the implementation of a specific (i-th) VR project,

$P_{CS\&ATP}$ (*profit, clarify standards and adjust technological parameters*) – profit of an industrial company received due to the clarification of standards and adjustments of technological parameters of various business processes due to the implementation of a specific (i-th) VR project,

P_{DNP} (*profit, develop new products*) – profit of an industrial company received due to the development of new products due to the implementation of a specific (i-th) VR project,

Table 6. Virtual reality projects implemented by major Russian ferrous metallurgy companies.

No.	No. From the Expert-400 rating	Company name, official website	Name of the VR-project	Essence of the VR-project
1	18	"EVRAZ", https://www.evraz.com/ru [29]	Virtual mine "My Mine"	The VR simulator is designed to train students in specialized mining specialties and improve the skills of staff
2	20	"NLMK", https://nlmk.com/ru [30]	Simulator of the unit "Ladle Furnace"	The VR simulator allows you to immerse yourself in the work of a steelmaker and control the steel production process in a simplified form
3	26	"Severstal", https://www.severstal.com/rus/ [31]	VR simulator	Teaching the steelworker's assistants on a VR simulator to take samples of liquid metal
4	29	"Metalloinvest", https://metalloinvest.com [32]	VR stand	Exposition with VR elements
5	30	"MMK", https://mmk.ru/ru [33]	1) VR simulator for training workers' safety rules 2) Modeling the technology of metallurgical production using a "digital" test site	1) Employees can take a virtual test on knowledge of labor protection rules at work 2) An own information system was created for predictive analytics of the quality of manufactured metallurgical products
6	108	"OMK", https://omk.ru/ [34]	System of remote repairs and installation of equipment in online mode	Using VR and AR technologies in combination with a "smart helmet", specialists at production sites perform diagnostics, repair and installation of equipment under the control of a remote operator

P_{ILP} (*profit, increase labour productivity*) – profit of an industrial company received due to the increase in labor productivity due to the implementation of a specific (i-th) VR project,

P_{RC} (*profit, reduce costs*) – profit of an industrial company received due to cost reduction due to the implementation of a specific (i-th) VR project,

P_{MM} (*profit, much more*) – profit of an industrial company, received due to various other factors (not taken into account earlier) due to the implementation of a specific (i-th) VR project.

3 Results

The authors believe that if there is a good basis for analytical accounting, as well as due to the digitalization of all business processes of an industrial company, it is possible to perform calculations using the above formulas, while comparing the economic feasibility of implementing different VR projects using standard formulas for calculating the economic efficiency of investment projects.

Thus, industrial companies that consider themselves innovative are actively implementing various VR projects in their activities, which leads to positive results: a reduction in operating costs, an increase in customer loyalty, an increase in income from the sale of goods, products, works and services, as well as strengthening the brand and business reputation of the company.

At the same time, it is possible to implement VR projects both by the company's own efforts and with the involvement of third-party developers of software solutions and mobile applications.

4 Discussion

Currently, only one scientific publication has been found in the Russian Science Citation Index database, which has the keywords "economic efficiency of VR-projects". It is devoted to discussing the use of virtual reality technology in construction as a way to save costs [35].

In Switzerland (in 2016), an article was published on a certain methodology for assessing the economic effect of the introduction of VR-projects, tested at industrial enterprises [36].

5 Conclusion

In general, companies are keen on using novel technologies like Virtual Reality (VR) in order to achieve competitive advantages. However, the economic impact of the integration of such technologies in a company is difficult to quantify.

It is necessary to further study the issues of evaluating various factors influencing the determination of economic efficiency from the implementation of VR projects by an industrial company in its business processes.

References

1. Virtual Reality Market Size (2021–2028). By Technology (Semi & Fully Immersive, Non-immersive), By Device (HMD, GTD), By Component (Hardware, Software), By Application, And Segment Forecasts. https://www.grandviewresearch.com/industry-analysis/virtual-reality-vr-market. Accessed 28 Dec 2022
2. Report on the research work "The use of virtual reality elements in the company's environmental business processes". Moscow, Financial University 99 (2022)
3. Official site Kopin Corporation. https://www.kopin.com/about/. Accessed 24 Dec 2022

4. Official site Lenovo Group Limited. https://www.lenovo.com/ru/ru/about/. Accessed 24 Dec 2022
5. Official site Samsung Electronics Co., Ltd. https://www.samsung.com/ru/. Accessed 24 Dec 2022
6. Official site Epson Corporation. https://global.epson.com/. Accessed 24 Dec 2022
7. Official site Sony Corporation. https://www.sony.com/en/. Accessed 24 Dec 2022
8. Official site Vuzix. https://www.vuzix.com/. Accessed 24 Dec 2022
9. Official site Apple. https://www.apple.com/ru/. Accessed 26 Dec 2022
10. Official site Blippar. https://www.blippar.com/. Accessed 26 Dec 2022
11. Official site Google. https://about.google/. Accessed 26 Dec 2022
12. Official site Magic Leap. https://www.magicleap.com/en-us. Accessed 26 Dec 2022
13. Official site Microsoft Corporation. https://www.microsoft.com/ru-ru. Accessed 26 Dec 2022
14. Official site Niantic Inc. https://nianticlabs.com/. Accessed 26 Dec 2022
15. Official site Wikitude. https://www.wikitude.com/. Accessed 26 Dec 2022
16. Official site Zappar. https://www.zappar.com/. Accessed 26 Dec 2022
17. Virtual Reality Headset Market Size, Share & Trends Analysis Report By End-device (Low-end, High-end), By Product Type (Standalone, Smartphone-enabled), By Application (Gaming, Education), And Segments Forecasts. https://www.grandviewresearch.com/industry-analysis/virtual-reality-vr-market. Accessed 27 Dec 2022
18. Official site Carl Zeiss AG. https://www.zeiss.ru/corporate/ru/home.html. Accessed 28 Dec 2022
19. Official site Facebook Technologies, LLC (Oculus). https://www.oculus.com/. Accessed 28 Dec 2022
20. Official site HTC Corporation. https://www.htc.com/eu/. Accessed 28 Dec 2022
21. Official site LG Electronics. https://www.lg.com/ru. Accessed 28 Dec 2022
22. Official site Razer. https://www.razer.ru/. Accessed 28 Dec 2022
23. Decree of the President of the Russian Federation of May 9, 2017 No. 203 "On the Strategy for the Development of the Information Society in the Russian Federation for 2017–2030"
24. Roadmap for the development of "end-to-end" digital technology "Technologies of virtual and augmented reality". https://digital.gov.ru/uploaded/files/07102019vrar.pdf. Accessed 10 Jan 2023
25. Expert-400: rating of the largest companies in Russia. https://expert.ru/expert400/2022/. Accessed 10 Jan 2023
26. Official site "Norilsk Nickel". https://www.nornickel.ru. Accessed 10 Jan 2023
27. Official site "Rusal". http://rusal.ru. Accessed 10 Jan 2023
28. Official site "UGMC Group". https://www.ugmk.com. Accessed 10 Jan 2023
29. Official site "EVRAZ". https://www.evraz.com/ru. Accessed 10 Jan 2023
30. Official site "NLMK". https://nlmk.com/ru. Accessed 10 Jan 2023
31. Official site "Severstal". https://www.severstal.com/rus/. Accessed 10 Jan 2023
32. Official site "Metalloinvest". https://metalloinvest.com. Accessed 10 Jan 2023
33. Official site "MMK". https://mmk.ru/ru. Accessed 10 Jan 2023
34. Official site "OMK". https://omk.ru/. Accessed 10 Jan 2023
35. Tolstikhina, E.D., Bryzhaty, D.R., Schmidt, M.I., Kudusov, A.A., Kremenskaya, E.A.: The use of virtual reality technology in construction as a way to save money (article). Econ. Sci. **190**, 84–89 (2020)
36. Ducker, J., Hafner, P., Ovtcharova, J.: Methodology for efficiency analysis of VR environments for industrial applications. In: International Conference on Augmented Reality, Virtual Reality and Computer Graphics, pp. 72–88 (2016)

Development of Numerical Methods for Reducing Stresses in Blades of Academic Turbine Under the Influence of Aerodynamic Loads

O. V. Repetckii and V. M. Nguyen(✉)

Irkutsk State Agrarian University named after A.A. Ezhevsky, Molodezhny, Irkutsk District, Irkutsk Region 664038, Russia
manhzhucov@gmail.com

Abstract. To reduce the vibration stresses that occur in the working blades of turbines during resonant excitations caused by the frequency of passage of the blades of the nozzle units, it is necessary to control the level of aerodynamic excitatory forces. One of the ways to reduce dynamic stresses in rotor blades under operating conditions close to resonant, in addition to structural damping, can be a decrease in external excitation forces. To reduce the intensity of the exciting forces, it is possible to use nozzle units with multi-pitch gratings, as well as with non-radially mounted blades of the nozzle units. The presented article shows the results of numerical calculations of exciting aerodynamic forces arising in academic rotor blades with a frequency zf, where f - is the rotor speed and z - is the number of nozzle blades [1]. The stage of the academic turbine was chosen as the object for research. Two options for changing the geometrical parameters of the turbine stage nozzle vanes are studied, namely: changing the angle of flow exit from the nozzle vane in the middle section and changing the profile of the nozzle vane. The presented results are obtained on the basis of numerical simulation of a viscous unsteady gas flow in a turbine stage using the ANSYS CFX computational gas dynamics software package, in which the numerical solution of the Reynolds-averaged Navier-Stokes equations is implemented.

Keywords: Aerodynamic Excitatory Forces · Numerical Calculations · Turbine Stage · Nozzle units · Working Blades

1 Introduction

With an increase in the unit power of steam and gas turbines, the requirements for operational reliability for the turbine unit as a whole and, in particular, for reliability for its most stressed elements - rotor blades have increased.

In addition to the constant forces of the steam flow that perform useful work, the rotor blades are affected by non-stationary aerodynamic forces that do not perform useful work, but from the point of view of the reliability of the rotor blades, they are the most dangerous [2].

© The Author(s), under exclusive license to Springer Nature Switzerland AG 2023
A. Gibadullin (Ed.): DITEM 2022, LNNS 683, pp. 112–121, 2023.
https://doi.org/10.1007/978-3-031-30926-7_11

Non-stationary aerodynamic forces that cause oscillations of the rotor blades, depending on the source of the disturbance, can be divided into two groups. The first group includes forces that have relatively low frequencies, multiples of the rotor speed. These forces are due to the circumferential non-uniformity of the flow, which in this case is associated with a deviation in the size and shape of the passage sections of the channels of the nozzle array, the presence of a partial supply and intermediate selections of the working fluid, etc. [3].

The second group includes aerodynamic forces caused by the presence of nozzle vanes, which create an uneven flow in pitch. The fundamental frequency of these forces is equal to the product of zf. At present, the following approaches can be considered as the most promising ways to reduce the amplitude of exciting forces: the use of nozzle array with a pitch variable along the circumference; non-radial setting of nozzle vanes; the choice of such design parameters of the flow path, in which the exciting forces are completely or partially extinguished, the change in the parameters of the nozzle array [4–6].

This article presents the results of calculations of aerodynamic loads applied to high-pressure turbine blades when changing the geometric parameters of the nozzle array stage.

2 Materials and Methods

With a variable pitch of the nozzle array, the excitatory forces have a frequency (kzf) (k $= 1, 2, 3\ldots$, f is the rotor speed, z is the number of nozzle blades). Since the gas-dynamic force F is a periodic value, it can be expanded in a Fourier series:

$$F = \sum_{k}^{\infty} F_k \cos(kzft - \gamma_k) \tag{1}$$

where F_k – the amplitude of the k component harmonic; k – harmonic number; γ_k– phase shift along the circumference.

In turbomachines, dynamic stress analysis is carried out only with the most dangerous harmonic (k = 1) [7]. Schematically the method of calculation is shown in Fig. 1.

In practice, it often becomes necessary to reduce the amplitudes of high-frequency resonant mode shapes. This problem becomes especially important when, for some reason, it is impossible to tune the rotor blades or when the turbomachine operates in a wide speed range. In such cases, intentional mistuning of nozzle gird pitches can be a fairly effective method for reducing dynamic stresses.

Ansys CFX 18.2 software package was used to calculate the three-dimensional viscous unsteady flow in the flow paths of the studied turbine stages. This package implements computational fluid dynamics (CFD) methods based on the use of non-stationary Reynolds-averaged Navier-Stokes equations (URANS). The Navier-Stokes equations for compressible flow are continuity, momentum, and energy equations in differential form. As a turbulence model, Menter SST model-one of the most popular turbulence model was chosen, which is suitable for calculating turbulent flows in turbomachines [8, 9].

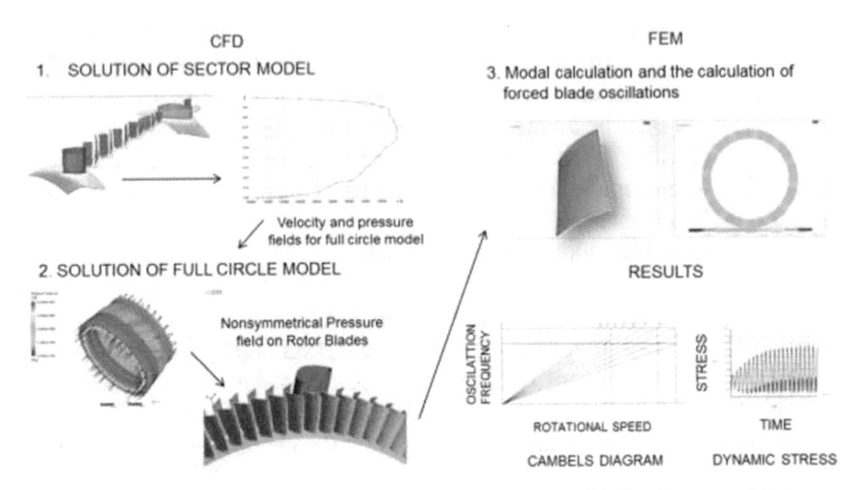

Fig. 1. Scheme of method for calculation compressor rotor blades forced oscillations.

An Ansys Turbogrid generator was used to construct computational grids for the blade sectors of the studied axial turbine stages. The total number of computational grid elements is 1.48 million. The dimensionless number y+ for the first near-wall cell for all grid models was in the range 1–2, which satisfies the requirements of the low Reynolds turbulence model SST.

The calculation time step was chosen so that the Courant number $C = \frac{u\Delta t}{\Delta x}$ (where u is the transfer rate, Δt- the time step, and Δx- the spatial step) was in the range of 1–4, in accordance with the requirements of the URANS method.

On the interface surface between the grids of guides and working interblade channels, the Transient Blade Row method with time transformation was used.

Research Objects. The aerodynamic profile of the impeller was chosen as the objects of study. On Fig. 2 shows the design model of the stage. The ratio of the number of guides and rotor blades was chosen so that the ratio of the angular sectors was close to one, in accordance with the requirements imposed by the Time Transformation method [10].

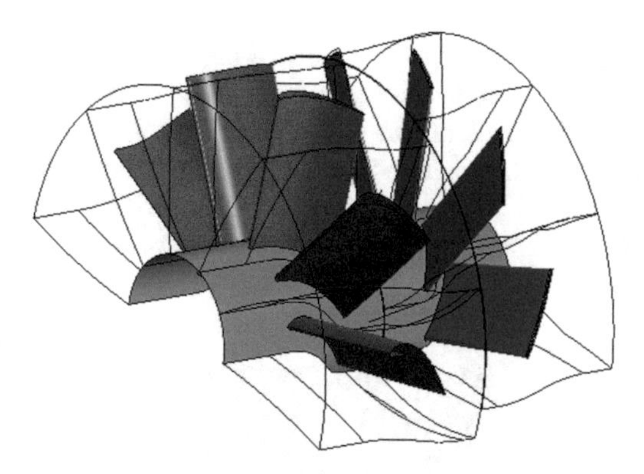

Fig. 2. Model stage.

The main geometric and operational characteristics of the studied stages are given in Table 1 [11]. The ratio of the total inlet pressure to the total outlet pressure reached $P_0/P_1 = 1.5$, the average mass temperature of the gas at the turbine inlet was 441.6 K. The Reynolds number calculated from the parameters at the outlet and the width of the rotor blades profile – $Re = 5.1 \times 10^5$. Rotor speed $\omega = 26.34$ 1/s is obtained from Campbell diagram and correspondingly with the most dangerous harmonics.

Table 1. Turbine stage geometry of the disk under study and boundary conditions.

Boundary conditions	Value
Number of nozzle vanes	10
Number of working blades	10
Passage height (inlet)	170 mm
Rotor speed	26.34 1/s
Pressure (inlet)	81,300 Pa
General temperature (inlet)	441.6 K
Average static pressure - (outlet)	53,250 Pa
Wall heat transfer	Adiabatic

3 Numerical Results and Discussion

In accordance with the objectives of this study, the influence of on variable aerodynamic forces acting on the rotor blades, the following geometric parameters:

- The angle of flow exit from the guide vane on the middle section;
- Type of nozzle vane profile.

Influence of the Value of the Angle of Flow Exit From the Guide Vane. The study of the influence of the angle of exit of the flow from the guide vane α_1 when determining the aerodynamic forces on the blade is always evaluated first. This is due to the fact that the intensity of unsteady processes is determined by the angle between the direction of the velocity vector of the flow incident on the rotor blades, which, in turn, depends on α_1. On Fig. 3 and Table 2 show the view and options for changing the angle of flow exit from the guide vane.

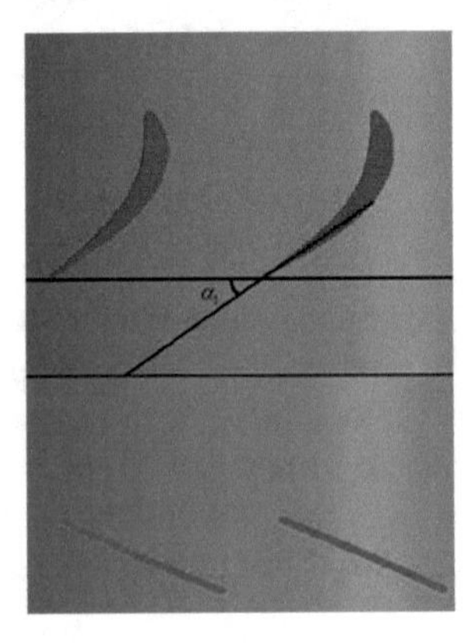

Fig. 3. Angle of flow exit from the guide vane at the middle section.

Table 2. Options for changing the angle of flow exit from the guide vane.

Option	Angle α_1
Option 1	$\alpha_1 = 40°$
Option 2	$\alpha_1 = 35°$
Option 3	$\alpha_1 = 30°$

After reaching the steady flow regime, for each variant, the calculation was carried out for three revolutions of the rotor with the recording of the values of the components of the force applied to the working blade. Plots of the ripple of the integral force over time for three options are shown in Fig. 4, the right side of the figure shows the time-average values of the force applied to the rotor blades: 274.4 N for option 1, 259.3 N for option 2 and 248.6 N for option 3. When the angle α_1 decreases by 5° the average value of the aerodynamic force decreases by 5.5%, and when the angle α_1 decreases by 10° the average value of the aerodynamic force decreases by 9.44%. Oscillation amplitudes of aerodynamic forces on rotor blades also decreases.

Further, calculations were made for the durability of the rotor blades under the action of aerodynamic loads obtained above. Figure 5 shows the results of calculations: the durability of the impeller is 43988 cycles at the angle of flow exit from the guide vane $\alpha_1 = 40°$; with a decrease in the angle by 5°, the life of the impeller increases by 17.41%, and with a decrease in the angle by 10°, the life of the impeller increases by 24.61%.

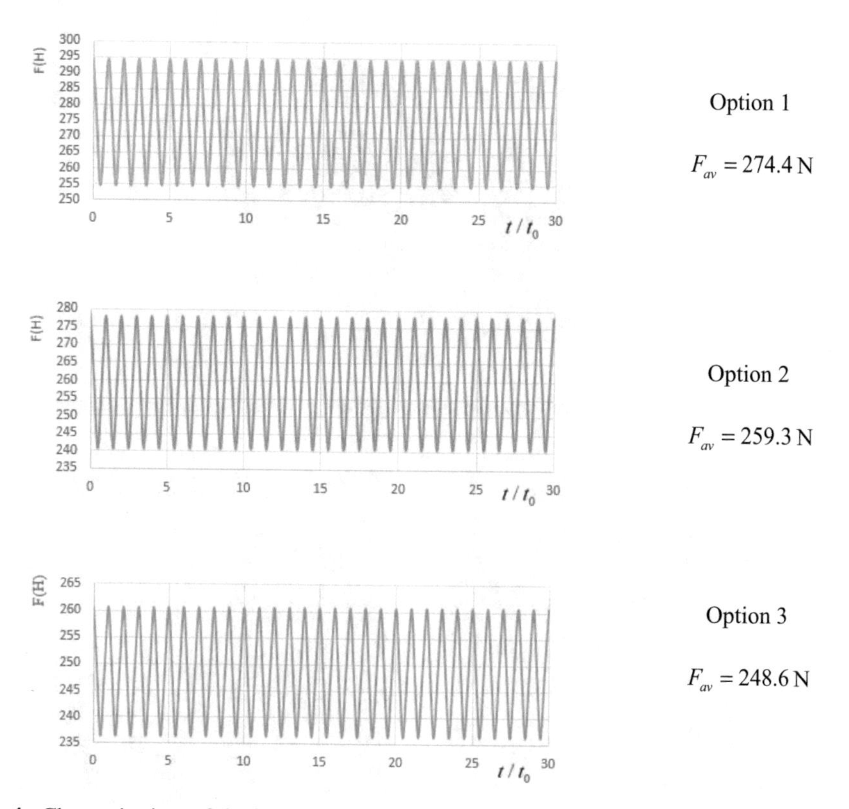

Fig. 4. Change in time of the integral force F acting on rotor blades, for variants 1, 2, 3 of the nozzle vanes geometry.

Changing the Profile of Nozzle Blades. At this stage, two variants of the nozzle blades profile with one angle of flow exit from the guide vane $\alpha_1 = 40°$ are considered (Fig. 6). Changes in the type of nozzle blades profile also have a great influence on the aerodynamic forces on the rotor blades, since with these changes the speed and direction of the oncoming flow on the rotor blades change.

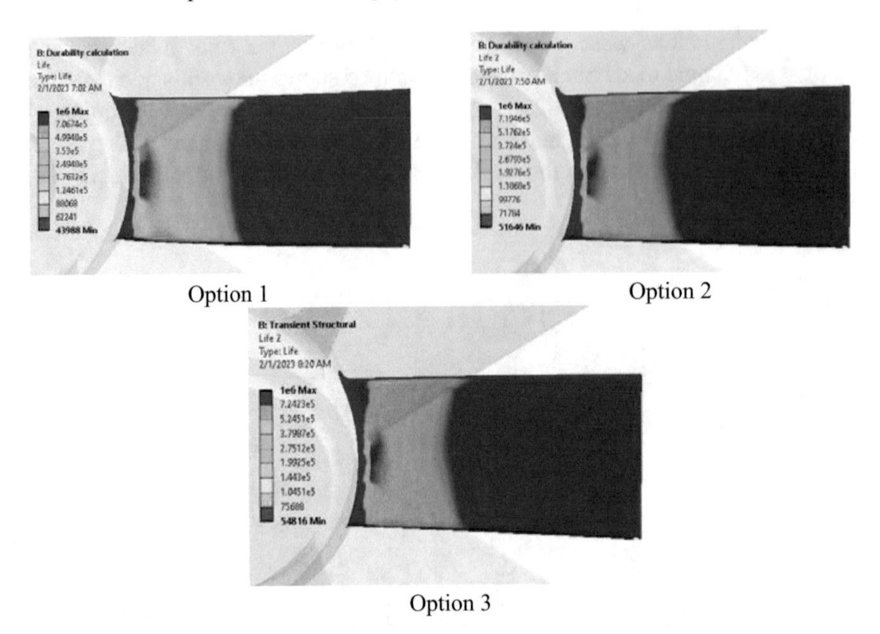

Fig. 5. Calculation of the durability of the impeller under the action of aerodynamic loads.

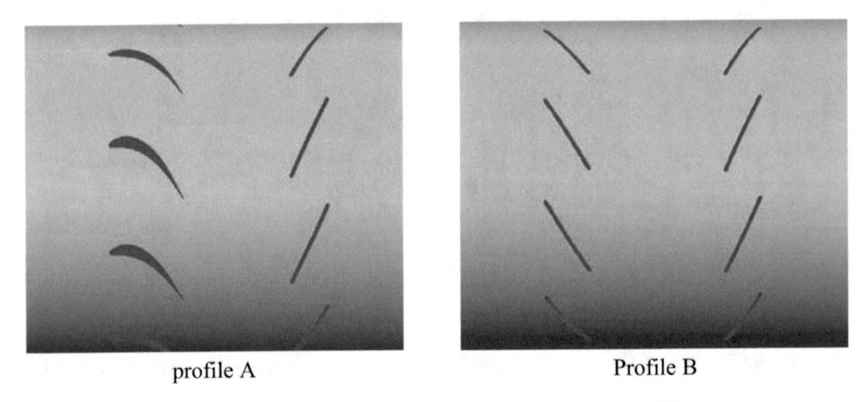

Fig. 6. Options for changing the nozzle vanes profile.

Figure 7 shows the diagrams of the pulsation of the integral force over time for two options for the value of the force applied to the blade: 248.6 N for option A and 228 N for option B. When using a profile of type B, the average value of the aerodynamic force is 8.29% less than when using a profile of type A.

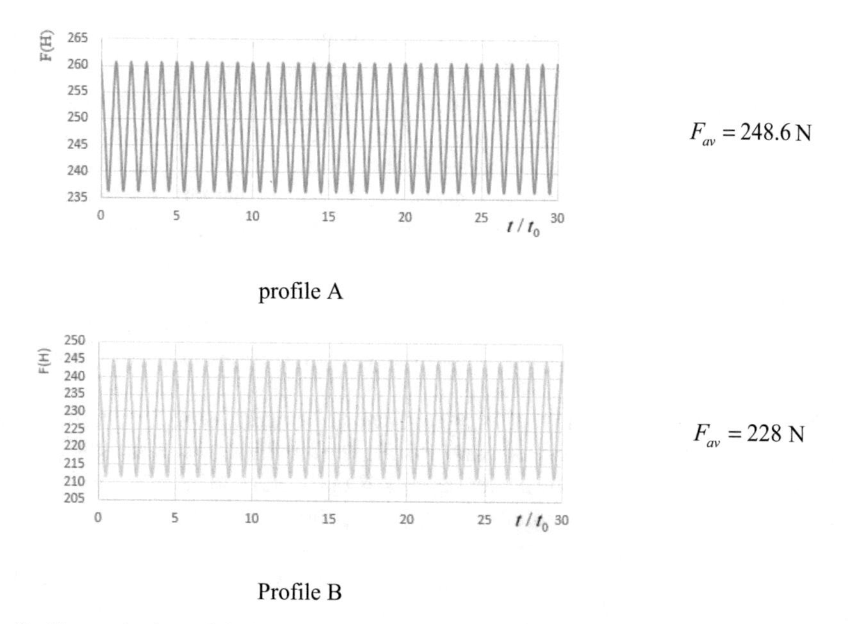

Fig. 7. Change in time of the integral force F acting on rotor blades, for variants A, B of nozzle vanes profile.

Also, calculations were made for the durability of the impeller under the action of aerodynamic loads (Fig. 8). When replacing a nozzle units with a section profile of type A for a nozzle units with a section profile of type B, the durability of the impeller increases from 54816 cycles to 57228 cycles, that is increases by 4.4%.

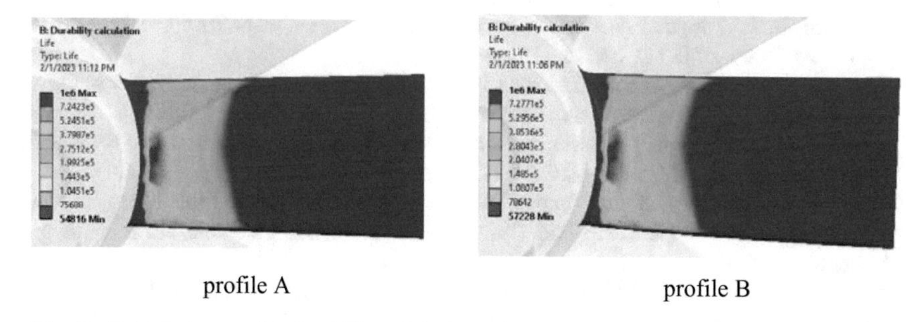

profile A profile B

Fig. 8. Calculation on the durability of the impeller under the action of aerodynamic loads.

4 Conclusions

Based on the results of a numerical analysis of the influence of geometrical parameters of nozzle vanes on the dynamic loads acting on high pressure turbine blades, it can be concluded that the results of studies of non-stationary processes in the flow parts of turbines can be used not only to solve the problems of ensuring the vibration reliability of the blade units, but also to assess the energy stage efficiency, as well as in solving problems of reducing aerodynamic loads on rotor blades. For a more complete simulation of non-stationary loading of turbine rotor blades, it is necessary to take into account all factors that affect the configuration of aerodynamic forces, namely: the unevenness of the temperature field behind the combustion chamber, the trace unevenness from the nozzle blades, and the influence of the geometrical parameters of nozzle vanes profile. The results of numerical calculations of this work show that a change in the angle of flow exit from the guide vane and a change in the type of nozzle vane profile can lead to a change in aerodynamic loads on the rotor blades, and therefore affect the durability of the impeller.

With a decrease in the angle of flow exit from the guide vane α_1 by 5°, the average value of the aerodynamic force decreases by 5.5%, the durability of the impeller increases by 17.41%. And when the angle decreases by 10° the average value of the aerodynamic force decreases by 9.44%, the durability of the impeller increases by 24.61%.

As a result of the replacement of the nozzle units with section profiles of type A to the nozzle units with section profiles of type B, a decrease in the aerodynamic excitation force by 8.29% was obtained by the calculation method, while the durability of the impeller increases by 4.4%.

The results of this numerical study on an academic impeller will be used to design of real gas turbine engines and serve as the basis for modeling the aerodynamic intentional mistuning of the working turbomachines stages [12].

References

1. Kostyuk, A.G., Frolov, V.V., Bulkin, A.Y.: Steam and Gas Turbines for Power Plants, pp. 452–473. MPEI Publishing House, Moscow (2016)
2. Repetskii, O.V.: Computer Analysis of the Dynamics and Strength of Turbomachines, 301 p. Publishing House of ISTU, Irkutsk (1999)
3. Rao, J.S., Rangarajan, N., Ratnakar, R.: Life calculation of first stage compressor blade of a trainer aircraft. ASME–Paper, vol. 7, 8 p. ASME Turbo Expo, Denmark (2012)
4. Kolenko, G.S., Laskin, A.S.: Unsteady and averaged aerodynamic loads acting on rotor blades of different geometry. Nat. Eng. Sci. **26**(1), 15–28 (2020)
5. Karaji, S.V., Tumashev, R.Z.: Comparison of aerodynamic characteristics blade rims with different blade axis shape. J. Sib. Fed. Univ. 245–257 (2012). Engineering and Technologies, Moscow
6. Kukhtin, Y.P., Shakalo, R.Y.: Reducing the vibration strength of pairs banded turbine blades bladis. Aviation and space technology i technology, no. 7, pp. 52–58, Zaporozhye (2020)
7. Popov, G., Kolmakova, D., Shklovets, A.: Optimization of the axial compressor flow passage to reduce the circumferential distortion. In: IOP Conference Series: Materials Science and Engineering, vol. 90, London (2015)
8. Dirk, W., Derek, M., Ronald, M.: Comparison of transient blade row methods for the CFD analysis of a high-pressure turbine. In: Proceedings of ASME, vol. 2D, 11 p. ASME Turbo Expo, Germany (2014)
9. Stuart, C., Mark, B.: A comparison of advanced numerical techniques to model transient flow in turbomachinery blade rows. ASME–Paper, vol. 7, 10 p. ASME Turbo Expo, Vancouver (2011)

10. Winhart, B., Micallef, D., Engelmann, D.: Application of the time transformation method for a detailed analysis of multistage blade row interactions in a shrouded turbine. In: Proceedings of 12th European Conference on Turbomachinery, 12 p. (2017)
11. Repetskii, O.V., Nguyen, V.M.: Application of methods for modeling aerodynamic forces on the rotor blades of turbomachines. In: Topical Issues of Agricultural Science, no. 43, 7 p. Irkutsk (2022)
12. Repetckii, O.V.: Digital and mathematical technologies in modeling the intentional aerodynamic mistuning of turbomachines. In: All-Russian Scientific and Practical Seminar: Digital Technologies in Science, Education and Production, pp. 52–53, Irkutsk (2022)

Growing Plant Cells Through the Integration of Additive and Information Technologies Using Statistical Methods

E. S. Bogodukhova[1], V. V. Britvina[1(✉)], A. V. Gavrilyuk[2], G. E. Nurgazina[3], and M. K. Pina[4]

[1] Moscow Polytechnic University, 38, St. Bolshaya Semyonovskaya, Moscow 107023, Russia
saaturn2015@mail.ru

[2] Lomonosov Moscow State University, 1, Leninskie Gory, Moscow 119991, Russia

[3] Academy of Intellectual Property, 55 a, Miklukho-Maklaya Street, Moscow, Russia

[4] Higher Institute of Educational Sciences of Huambo-Angola, Avenue Alioune Blondin Beye, Academic District, Huambo 2376, Angola

Abstract. This work is devoted to the integration of advanced technologies in the agricultural industry into crop production for growing plants from plant cells by adhesion in a special environment. The aim of the work is to create a technology for growing living organisms through a combination of additive and information technologies. An empirical analysis of the world's biosystems has shown that the reduction in plant species diversity is interconnected with the climatic crisis that is progressing today. The study of plants in various conditions of existence made it possible to assess the ecological situation in the world and identify the main directions for eliminating negative consequences. To prevent the destruction of ecosystems, an alternative form of growing vegetation of natural origin has been proposed, which will speed up the process of growing plants without soil, thereby not draining it. The data presented, confirming the practical importance of additive technologies for the development of crop production and bioengineering, are aimed at reducing the amount of CO_2 emissions into the atmosphere, stabilizing climate change and improving the climate situation. The article is supplied with graphic materials and tables, as well as a detailed description of each stage of the study.

Keywords: Additive Technologies · Information Technologies · Statistical Methods · Plant Cells

1 Introduction

Recently, visible changes have been taking place in biosystems around the world: climatic conditions destroy the microstructure of natural origin, significantly reducing the species diversity of wild and cultivated plants, and the number of plant groups included in the third edition of the Red Book of the Russian Federation, which has increased almost 1.5 times, led to a decrease in primary production in ecosystems. The imbalance of all natural

© The Author(s), under exclusive license to Springer Nature Switzerland AG 2023
A. Gibadullin (Ed.): DITEM 2022, LNNS 683, pp. 122–132, 2023.
https://doi.org/10.1007/978-3-031-30926-7_12

systems associated with the observed climate crisis [1] led to irreversible consequences and caused colossal damage to the entire biosphere.

According to scientists, the extinction of biological species continues at an accelerating pace, and therefore, emergency measures are needed to help prevent an impending global ecological catastrophe. Over the past 50 years, humankind has been destroying natural ecosystems and in the coming decades, about 45% of the planet's animal and plant species are under threat [8].

That is why, in order to prevent negative consequences, it is necessary to consider alternative types of growing vegetation of natural origin, the purpose of which is to update bioprinting using plant tissue cells to obtain plants of specified shapes and sizes. To interpret positive trends, it is necessary to study the extent of the imbalance, identify the growth potential of alternative growing of plant cells [10], and determine the significance of such projects in the Russian Federation.

2 Materials and Methods

Here is how to display a pop-up window from which to select and apply the AIP Conference Proceedings template paragraph styles. To assess the scale of changes occurring with a decrease in species diversity, we used databases containing the results of observations and experiments on the impact of factors such as climate change [2], an increase in the concentration of carbon dioxide, an increase in the concentration of nitrogen, phosphorus and other biogenic elements available to plants in water and soil, an increase in the acidity of the environment and others. The data include the results of observations and experiments in terrestrial, freshwater and marine conditions [3].

Based on the study, it was found that a moderate decrease in species diversity within the range of 20–40% leads to a decrease in the primary production of plants by 5–10%. A twofold decrease in species diversity leads to a decrease in the primary production of ecosystems by an average of 13%. Higher levels, where a decrease in diversity within 41–60% of the original, in terms of influence on the production process, are comparable to the effects of such global factors as eutrophication, increased acidity, or an increase in the concentration of carbon dioxide (see Fig. 1) [3].

The logarithm of the ratio of experimental productivity to product control is a measure for evaluating the positive and negative effects under given conditions [4]. According to the data presented in Fig. 1, the negative effect arising from various globally acting factors with a decrease in plant species diversity will be 95%, which will lead to an increase in CO_2 concentration and the intensity of ultraviolet radiation, as well as to climate warming. With a further depletion of the composition of destructors, the rate of decomposition of plant residues decreases by about 20% (see Fig. 2) [4]. Accordingly, there is a depletion of soil ecosystems that perform ecological functions and are regulators of the content of CO_2, N_2, O_2 in the air, as well as absorbers of harmful gas impurities [5], the soil formers of which slow down the destruction process and have a significant impact on the full existence of ecosystems (see Fig. 2).

To prevent this plant effect, it is necessary to use technologies capable of growing living cells in artificial nutrient media [6]. Cultivated cells react strongly to changes in the pH of the medium, so the pH value of most cells is in the range of 7.2–7.4. Sodium

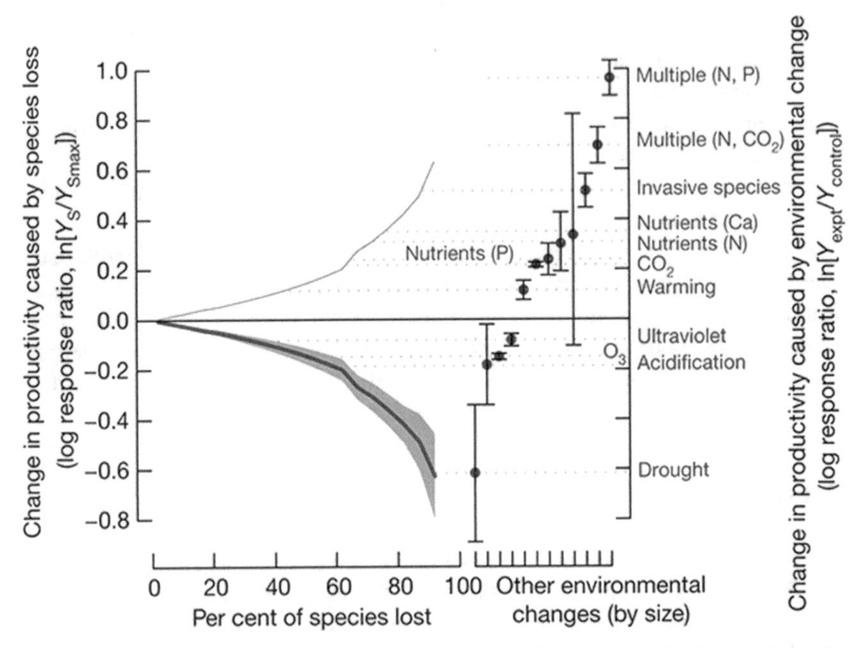

Fig. 1. Changes in primary production in ecosystems with a decrease in the number of plant species.

bicarbonate maintains the "natural" buffer system of the culture medium (CO32-/HCO3-), requiring 5–10% CO_2 in the atmosphere, which is easily accomplished when cells are cultured in a CO_2 incubator. Hepes is a phosphate salt with a buffering capacity in the range of 7.2–7.4 pH. It does not require high pressure in a gaseous environment, but in high concentrations, it can be toxic to some cell types. Phenol red is used as a pH indicator: red at pH 7.4 and changes to orange or yellow when the pH value is measured. For the cultivation of estrogen sensitive cells, it is recommended to use a medium without phenol red. Inorganic salts of the total osmotic pressure in the middle and middle regulation of the diaphragmatic space, perceiving the absorption of sodium, potassium ions and concentrations. The osmolality of the medium is as important as the pH level in cell culture. The optimal value of osmolality lies in the range of 260–340 mosmol/kg, depending on the type of cultured cells. Amino acids are the building blocks of proteins; essential amino acids are always included in the culture medium. L-glutamine is especially important; provides NAD, NADPH and nucleotides with nitrogen, being a secondary source of energy for cell metabolism. Non-essential amino acids are also sometimes added to the culture medium. Also, most media include glucose and galactose as an energy source for cells, and the most commonly used proteins and peptides are albumin, transferrin, and fibronectin; they are especially important in serum-free cultivation. Albumin binds water, salts, hormones and vitamins and transports them between cells and tissues. Fibronectin plays a key role in cell adhesion. Transferrin is an iron transfer protein that provides iron to the cell membrane, especially important in serum-free culture, since serum usually contains them.

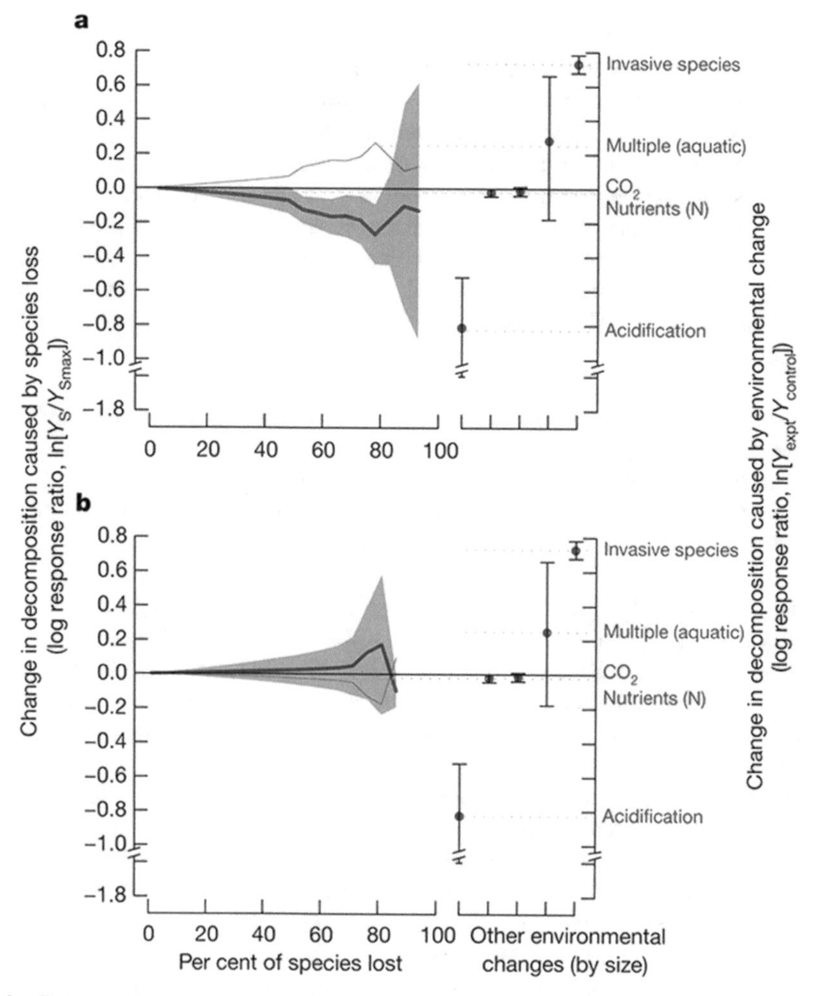

Fig. 2. Change in the rate of destruction of organic matter with a decrease in species diversity: a – destructors; b – producens.

Many vitamins are essential for cell growth and proliferation, so cultured intake typically includes riboflavin, thiamine, and biotin. The most important component of the culture medium is whey; it was made before use, approximately 5–10%. Serum is a mixture of albumins, growth factors and growth inhibitors; is a source of nutrition for vitamins, amino acids, proteins, carbohydrates, carbohydrates, trace elements, growth products. Most commonly used is normal and fetal calf serum. Growth factors, cytokines, hormones are enhanced in the culture medium for cell proliferation and activation. Antibiotics are added to prevent contamination of the culture medium by bacteria and fungi; however, antibiotics do not prevent infection of the culture medium with mycoplasma [11]. Thus, 3D bioprinting helps to grow plants without a natural environment using additive technologies, creating the necessary conditions.

However, no less rapid growth in growing plants is cell division. The preparations produce phytohormones necessary for cell division under special conditions, which consume low-molecular compounds in very low concentrations, performing regulatory functions and capable of significantly influencing processes such as the germination of tubers and seeds, the growth and division of tissues, flowering, and the formation of fruits. But such environmental factors do not always add up, therefore, a method was developed for obtaining phytohormones from external sources, which explores itself with the help of specific top dressings [12], due to the fact that plants may not increase productivity, but also abundant fruiting in plants under the influence of such complexes, among which several main groups of compounds associated with phytohormones can be found:

- Auxins are hormones concentrated in the upper sections of the stems, their main effect is the stretching of plant cells;
- Gibberellins - concentrated in the leaves, characteristic of cell growth;
- Cytokinins - penetrate into the tips of the roots and seeds; regulate the processes of cell division.

Today, for the cultivation of various biological objects, including microorganisms and cell cultures, depending on the tasks set, various technological equipment is used. These can be rocking rollers - in laboratory research, a bioreactor and fermenters - in industrial production. Industrial cultivation of plant cells is carried out in bioreactors or fermenters. They are divided into two groups: by design and by the principle of mixing the culture fluid. In bioreactors belonging to the first group, cells are mixed by aeration with air. This is a bubbling type of bioreactor, in which the process of mixing the suspension is carried out by rising air bubbles. In the case of bubble bioreactors, good growth characteristics are usually obtained for a large number of cell cultures. However, the difficulty of maintaining a suspension in a homogeneous state at high concentrations of cell biomass narrows the scope of their application. Several higher values of the maximum concentration of cell biomass can be achieved using airlift bioreactors, in which directed circulation flows are created. In airlift bioreactors, the mixing of the suspension is carried out by using a special design that creates a density gradient (as a rule, this is a design with an internal cylinder) [13].

The second group of bioreactors are apparatuses using mechanical agitators. Bioreactors of this type make it possible to study plant cell populations in a very wide range of cell biomass concentrations. At the same time, the stress effect of the mixing device on the cell population often limits their application [13].

With regards, to additive technologies, they allow the generation of cellular structures similar to the growth of silicone in a microgel to release drugs and then control tissue regeneration. The silicone model is printed on a micro-organic gel backing. The printing nozzle follows a predetermined trajectory, squeezing out the liquid silicone. The 3D printing method is engaged in the manufacture of cellular structures up to hours. Different interpretations of 3D printing for growing plants: from creating a user-selected shape with plant seeds inside, which in a few weeks turn into a full-fledged garden [14], to micro growing fibrous structures, which, during adhesion, become a multicellular organism of plant origin, in the near future will allow artificial creation biological organisms of flora and fauna in the shortest possible time, which will not only stop the extinction of various plant species, but also prevent a global environmental catastrophe.

At the moment, the structure of the market of additive technologies in percent by areas of energy, aviation and space, industry, medicine and other goods, including energy, food, construction and bioprinting, the percentage of which is less than one, so the cultivation of natural tissues is still underdeveloped and has huge potential for further development.

3 Results

Based on the data obtained in the course of the study, it can be confirmed that this technology of layered cultivation of crops is relevant for many countries; it combines the development of three-dimensional tissue modeling and special additive equipment based on FDM-printing. The first stage is the construction of the plant cell and the construction of a three-dimensional drawing, based on which a file in STL format is created. The digital model contains information about the relative position of the points, after which the model is processed in a special program. The computerized process of layer-by-layer bioprinting takes place directly with the selection of a suitable medium in the bioreactor for conducting experiments and monitoring the process of cell maturation.

Cells of future plants are placed in an artificial nutrient medium for growth and/or proliferation. The main factors required for various cell growth are elevated temperature, cell attachment substrate (for cell attachment to cells), suitable culture medium, CO_2 incubator to control the pH level, use of osmotic. The most important point to ensure the overall growth/proliferation of cells is the choice of a suitable culture medium. The culture medium is a gel observed for cell uptake/proliferation growth. The cell culture medium consists of certain amino acids, vitamins, salts, glucose, hormones, nutritional supplements, growth carbohydrates, supporting buffer system and osmotic pressure [11].

Since the basis of plant growth is cell division, therefore, in the process of growing a plant using 3D printing, specific additives are provided that can accelerate the process of cell division, stimulating their development. They contain phytohormones - organic substances, on the action of which the processes of growth and development of plants directly depend. Such substances differ in their chemical structure, the degree of influence on biological organisms, and, accordingly, the results [12]. Depending on whether these substances are manifested, the functions of the regulators are activated. It is important that such dressings also include not only hormones, but also other components, essential plants, in natural proportions. This is their proper nutrition, rapid growth and a high probability of stress, stimulation of development. The most relevant application of the stimulant formula with a visible balance is for growing for this work a complex of phytohormones, humic and fulvic acids, monosaccharides, macro- and mesoelements, as well as amino acids and fatty acids.

Within the framework of this work, a 3 D-printing biological head was designed for layer-by-layer growing of living cells, which is an optimized design of a classic 3D printing head for FDM-printing with a micro-screw. A special working body allows not only moving living cells along, but also the function of a nozzle that forms each layer. The main components of the 3D microprinted head are NEMA17 stepper motor, friction clutch, hardened steel screw and cylinder with heat sink, aluminum thermal block, thermocouple, fans, nozzle, and needle and feed hopper. The materials were obtained depending on their function: hardened steel retains high hardness and wear resistance when printing, copper and aluminum have high thermal conductivity. After that, the geometric parameters of the 3D printing head were calculated, where all calculations were carried out according to the methodological manual of Yu I Litvinets. "Technological and energy calculations in the processing of polymers by extrusion" [7] according to the formulas indicated in it. Thus, the geometrical parameters of the screw were calculated (see Table 1), the parameters of the 3D printing head (see Table 2), and the performance of the microscrew head (see Table 3).

Table 1. Calculation of the geometric parameters of the extruder.

Parameter	Designation	Meaning	Units	Formula	Note
Volumetric productivity	Q	4	mm^3/s	$Q = 0,6D^{2,5}$	
Screw diameter	D	2	mm		
Screw pitch	t	1.6	mm	$t = k_1 \cdot D$	$k_1 = 0.8$
Screw length	L	16	mm	$L = k_2 \cdot D$	$k_2 = 10.3$
Screw length to compression zone	L_0	6.4	mm	$L_0 = 0,4L$	
Screw head length	L_H	9.6	mm	$L_H = 0,6L$	
Channel depth in the feed zone	h_1	0.24	mm	$h_1 = k_3 \cdot D$	$k_3 = 0.12$
In the plasticization zone	h2	0.2	mm	$h2 = h1 - \frac{h1-h3}{L} \cdot L_0$	
In the dosing area	h3	0.1	mm	$h3 = 0,5\left[D - \sqrt{D^2 - \frac{4h1}{i}(D - h1)}\right]$ i = 2	i = 2
Coil crest width	e	0.4	mm	$e = k_4 \cdot D$	$k_4 = 0.1$
Radial clearance	δ	0.010	mm	$\delta = k_5 \cdot D$	$k_5 = 0.001$

Table 2. 3D print head operation parameters.

Parameter	Designation	Meaning	Units	Formula	Note
Critical speed	n_{un}	0,31	s^{-1}	$n_{un} = \frac{42,2}{60\sqrt{D}}$	
Operating speed*	n_p	0,22	s^{-1}	$n_p = k_6 \cdot n_{un}$	$k_6 = 0.65$

*In practice, the rotational speed is in the range from 0.08 to 4.2 $s^{(-1)}$. Recently, adiabatic (autothermal) screw machines with rotation speeds up to 8.4 $s^{(-1)}$ have begun to be used.

Table 3. Screw performance calculation.

Parameter	Designation	Meaning	Units	Formula
Volumetric performance	Q	0.92	cm^3/min	$Q = \frac{A_1 K}{K+B_1+C_1} n$
Head drag coefficient	K	0.000199	cm^3	$\frac{\pi d^4}{128L}$
Screw speed	n	13.21	$min^{(-1)}$	
Constants of the forward and reverse flows and the flow of leaks, respectively, for a screw with a variable depth of thread	A_1	6.85945		$A_1 = \frac{\pi^3(t-\lambda e)\sigma}{a+t^2 b}$
	B_1	0.00013		$B_1 = \frac{\pi t(t-\lambda e)}{12L_H(a+t^2 b)}$
	C_1	0		$C_1 = \frac{\pi D\delta^3 t^2}{10eL_H\sqrt{\pi^2 D^2+t^2}}$
Coefficients characterizing the design of the screw with a variable depth of cut	σ	75.34		$\sigma = 1 - \frac{6,9D}{2(h_2-h_3)} \log \frac{h_2}{h_3} + \frac{D^2}{2h_2 h_3}$
	a	989.20		$a = \frac{\pi^2}{h_2 h_3}\left[\frac{D(h_2+h_3)}{2h_2 h_3} - 1\right]$
	b	1.93		$b = \frac{2,3}{(h_2-h_3)D^3} \log \frac{h_2(D+d_3)}{h_3(D+d_1)} + \frac{2h_2 h_3+(h_2+h_3)D}{2D^2 h_2^2 h_3^2}$
Core diameter in the loading area	d_1	3.8	mm	$d_1 = D - 2h_1$
Core diameter in the ozing zone	d_3	4.4	mm	$d_3 = D - 2h_3$
Nozzle diameter	d	0.1	mm	
Nozzle channel width	L	1	mm	

After calculating all the parameters, a stepper motor was selected by power "NEMA17HS4401" with the following characteristics, presented in Table 4.

In addition, after carrying out all the necessary calculations, a parametric 3D model of an improved 3D printing head with a medium for growing plant cells was designed

Table 4. Stepper motor power calculation.

Parameter	Designation	Meaning	Units
Rated operational current	I	1.7	A
Winding resistance	R	1.5	Om
Moment of inertia	I_{in}	54	g/cm^2
Torque	M	40	Nm
Voltage	U	24	Volt
Engine power 17HS4401	N_1	40.8	Tue
Shaft power	N_2	40.6	Tue

(see Fig. 3). The work of a 3D-printing bio head is carried out as follows: the coupling transmits rotation with a stepper motor to the auger, which in turn moves the plant cells [9] into a special nozzle that looks like a thin needle with a hole in the middle, with the help of which the plants are grown layer-by-layer. The whole process is carried out in such an environment for a certain type of plant, which improves the adhesion of plant cells when they are 3 D-printed. Similar to a bioreactor, the cell mixing process takes place in the cylinder to obtain good growth characteristics for most cell cultures. The 3D print head growth process differs from industrial plants in that it maintains the suspension in a homogeneous state at high concentrations of cell biomass by creating a density gradient. The proposed cultivation technology is relevant not only for agronomy and plants, but also for those areas where living cells are used as a consumable material, which must be grown quickly and efficiently, not forgetting about the medium for adhesion of cell material, for example, in bioengineering.

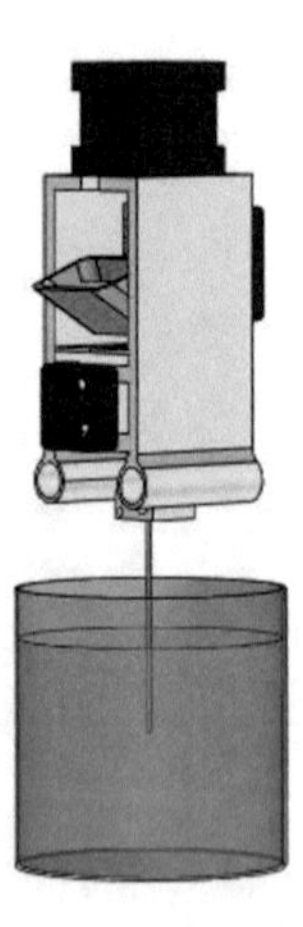

Fig. 3. 3D model of the advanced print head.

4 Discussion

Based on the study, it can be noted that the proposed technology will prevent a decrease in species diversity, and, therefore, will lead to a balance of the primary production of the ecosystem, which will reduce the acidity and concentration of carbon dioxide. The negative effect (see Fig. 1) will be reduced to 50%, thereby not having a strong impact on climate change (see Fig. 2), performing environmental functions to the fullest and absorbing harmful gases of impurities. The generation of certain types of biological organisms of flora and fauna will allow not only to grow the required shoots, but also to restore damaged plants by growing cells on their surface by diffusion penetration of cellular structures. The directly proportional dependence of the beneficial effect of the presented technology on the increase in the number of plants will allow reaching the natural balance as soon as possible, thereby reducing the percentage of the impending climate threat.

A detailed study of existing nutrient media made it possible to choose the most effective medium formula for layer-by-layer three-dimensional plant growth, and environmental factors confirmed the need for the use of phytohormones, which will be a kind of special type of top dressing.

An analysis of existing systems for the cultivation of various biological objects, including microorganisms and cell cultures, determined the direction of development for updating and proved that there is no technology for growing such 3D printing today. This is confirmed by the absence of foreign and domestic developments of a 3D printed biological head, which proves the practical significance of the project for further studying the possibilities of growing various plant species, not only on the verge of extinction, but also hybridization of new unique properties that will help stabilize the climate situation.

The creation of a 3D model of a plant growing system made it possible to clearly demonstrate how the construction process will take place, and displayed all the constituent elements of the print head, an important organ of which is a needle for the formation of plant cells, which directly prints after immersion in a nutrient medium. Such a consistent movement of the construction cells will allow maintaining a homogeneous state throughout the entire printing process, regardless of the concentration of cell biomass, as well as breeding new plant species due to the centralized application of plant material according to the construction points.

Summing up the work done, it can be noted that the artificial plant cultivation project is an integration of advanced technologies, in which the positive effect is achieved much faster and better with minimal investment and time loss, reducing the cycle of growing plant cells for instant herbal remedies, its printing proves it economic feasibility.

The data obtained as a result of the information search confirmed the importance of conducting research work on the manufacture of a micro screw print head for 3D printing with plant cells in a special nutrient medium. The main task of R&D is to study the modes of operation of a 3D printing head. The result of the study will be the establishment of a rational mode of operation of the 3D printer. Future research on testing the growing system will determine the future direction of the project, which is the subject of future publications and patents for inventions in the field of additive and biotechnologies.

5 Conclusion

In conclusion, I would like to add that the bioprinting market is far from being oversaturated and is innovative in many areas, where the integration of additive technologies with both advanced and traditional technologies can have great potential for the dynamic development of both agriculture and other industries.

References

1. Bogodukhova, E.S.: IOP Conf. Series: Earth Environ. Sci. 723, 052034 (2021)
2. Bogodukhova, E.S.: IOP Conf. Series: Earth Environ. Sci. 808, 012009 (2021)
3. Decrease in species diversity in ecosystem. https://elementy.ru/novosti_nauki/431864/Umenshenie_vidovogo_raznoobraziya_privodit_k_snizheniyu_pervichnoy_produktsii_v_ekosistemak. Accessed 16 Oct 2022
4. Logachev, M.S., Voronin, I.V., Britvina, V.V., Tichtchenko, S.A., Altoukhov, A.V.: Local area network monitoring: the issue of broadcast storm. Int. J. Adv. Trends Comput. Sci. Eng. **9**, 4216–4222 (2020)
5. Conservation and restoration of biodiversity. http://www.nature.air.ru/biodiversity/book2_4.html. Accessed 16 Oct 2022
6. Ditchenko, T. I.: Culture of cells, fabrics and plant organs. BSU (2007)
7. Litvinets, Y.I.: Tekhnologicheskie is energeticheskie raschety pri pererabotke polymers by extrusion: method. instructions for practical. classes, coursework and diploma. design of specialty 240502 "Technology of processing of plastics and elastomers". Yekaterinburg, UGLTU (2010)
8. Formation of the climate crisis. https://climate.greenpeace.ru/. Accessed 16 Oct 2022
9. 3d-model biology plant cell. https://sketchfab.com/3d-models/biology-plant-cell-0d87fbe2581b4cabae809bd6aaa7d56b. Accessed 17 Oct 2022
10. Glick, B., Parsnip, J.: Molecular Biotechnology. Principles and Application. Mir, 589 p. (2002)
11. Cell culture media. https://www.dia-m.ru/page/sredy-dlya-kultivirovaniya-kletok/. Accessed 18 Oct 2022
12. Cell division stimulators. https://lignohumate.ru/catalog-gumatov/stimulyatory-deleniya-kletok/. Accessed 19 Oct 2022
13. Vavilov, N.I.: Biotechnology for the production of proteins and biologically active substances: a short course of lectures for students of the 1st year of training 19.04.01 "Biotechnology". FGBOU VO Saratov State Agrarian University, 87 p. (2016)
14. 3D printed green gardens. https://www.facepla.net/the-news/tech-news-mnu/4921-3d-printer-for-green-trees.html. Accessed 19 Oct 2022

Automated Address Storage and Accounting of Spare Parts at Railway Transport Facilities

A. V. Klyukanov[1], A. L. Zolkin[2](✉), O. V. Saradzheva[3], V. V. Dragulenko[4], and A. S. Bityutskiy[5]

[1] "Technologies of Freight and Commercial Operations, Stations and Centres" Department, Samara State Transport University (SSTU), 2V, Svobody Street, Samara 443066, Russia

[2] Computer and Information Sciences Department, Povolzhskiy State University of Telecommunications and Informatics, Samara 443010, Russia
alzolkin@list.ru

[3] "Invent Economic Security, Controlling and Audit" Department, Russian State University Named After A.N. Kosygin, 33, Sadovnicheskaya Street, Moscow 115035, Russia

[4] Department of Tractors, Automobiles and Technical Mechanics, Federal State Budgetary Educational Institution of Higher Education, "Kuban State Agrarian University Named After I.T. Trubilin", 13, Kalinina Street, Krasnodar 350044, Russia

[5] "Invent Technology" LLP, Almaty A10E5P4, Kazakhstan

Abstract. An automated system for accounting of the availability and consumption of spare parts at railway transport facilities is considered in this study. The system includes racks with cells placed in closed cabinets and integrated into an information network connected to a remote terminal for collection, processing and transmitting of information. On the racks there are containers with products and an information input panel containing a microcontroller connected via a data transmission device to a remote terminal, a keyboard and a display, as well as an operator's electronic key. The technical features of the proposed automated system is the presence of additional tensometry sensors with the required reference value of the state of the rack cell. The system also includes a switching unit, an analog-to-digital converter, a cell address generation unit, a communication unit and a mobile receiver with output of information about the location of the completed nearby racks, the number and type of spare part. The cell state sensors are connected to the information inputs of the switching unit, the output of which is connected to the input of the analog-to-digital converter. The proposed automated system for accounting of the availability and consumption of spare parts at railway transport facilities provides a prompt search for completed racks and storage cells, increasing the information content of accounting of the availability and consumption of spare parts in racks, accumulating information about the number and type of storage elements for each rack included in the automated complex, as well as the ability to change the number of cells for storage of the same type of parts, depending on the load on the racks.

Keywords: Rack · Automated System · Address Storage · Tensometry · Information Input · Electronic Key · Information Network

© The Author(s), under exclusive license to Springer Nature Switzerland AG 2023
A. Gibadullin (Ed.): DITEM 2022, LNNS 683, pp. 133–140, 2023.
https://doi.org/10.1007/978-3-031-30926-7_13

1 Introduction

Automated address storage and accounting of spare parts is used in the warehousing of many industries, including railway transport, but to a lesser extent. For example, electronic racks for storage and automatic accounting of spare parts developed by the Design and Technology Institute of Scientific Instrumentation of the Siberian Branch of the Russian Academy of Sciences are in trial operation in the Inskaya railroad car shed [2, 4].

Automated racks differ in design, overall dimensions, electronic filling of control sensors. At the same time, the principle of operation of all such racks is usually based on tensometry [1]. The same type of parts is placed on the shelves of the rack place. When the consumables on the racks fall below the permissible standard values, the pressure on the shelves decreases due to gravitational forces and a sensor is triggered. It opens the contactor circuit. After that, the electrical signal enters the automated control system for the storage and accounting of spare parts through the coupling, through the communication channels. Than here it is converted into a quantitative indicator of the parts on the rack. The rack can also be equipped with a light indication, which informs the worker about its' fullness with spare parts. Information about the replenishment of the rack is automatically transferred to the electronic system of the warehouse from the system of accounting and storage of spare parts.

The disadvantage of such an automated rack is that the information accumulated in the automated spare parts storage and accounting management system is used by the operator mainly to replenish the racks with components through the warehouse workers (storekeeper, vehicle driver). In the conditions of operational work, it can lead to interruptions in the provision of wagons with serviceable spare parts.

Also, the worker can receive information about the availability of spare parts on the racks from the warehouse operator by telephone or other communication. At the same time, the operator will need time to respond to the worker's request, which will lead to unproductive loss of working time.

2 Problem Statement

In this study, it is proposed to expand the functionality of automated address storage and accounting of spare parts in warehousing.

The technical result is achieved by the fact that an automated system for accounting of the availability and consumption of spare parts, consisting of racks with cells equipped with storage tensometry, integrated into an information network, transmit information to a remote terminal [3, 7]. The input device consists of a display, a keyboard for displaying the state of the cells, a transmitter, additionally introduced removable cells with tensometry sensors with the required setting value, a receiving device with storage memory, an operating unit with a rack selection button, an electromechanical lock of the rack cells with one electronic reader, a rack accounting and load analysis unit, a mobile receiving device with the output of information about the location of the nearest completed racks, the number and type of part [5, 6].

3 Research Questions

In an automated system for accounting of the availability and consumption of spare parts, electronic racks are an integral part. At present, the widespread use of automated racks is hindered by the high cost of control sensors and electrical communications. With development of scientific and technological progress (the emergence of new materials, other principles of signal transmission), the cost of an automated rack in the system for accounting and consumption of spare parts will decrease.

4 Materials and Methods

According to the existing technology, the racks are replenished from the warehouse with the delivery of spare parts to the racks on mobile vehicles without taking into account the range of used parts (Fig. 1). Due to the different intervals for the arrival of freight cars on the fleet tracks, the maintenance work on these tracks is uneven. In addition, the failures of the wagon units are random, which is also expressed by the different laboriousness of uncoupled repairs on the fleet tracks.

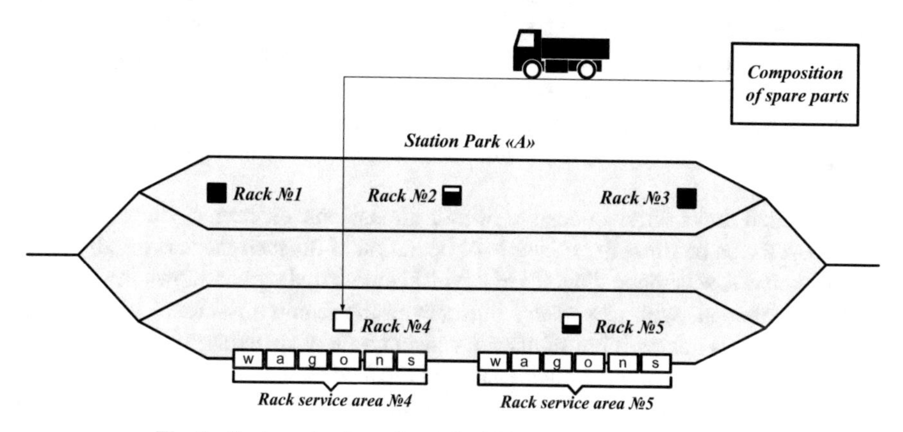

Fig. 1. Basic technology for replenishing racks with spare parts.

All this leads to the fact that on the tracks, which are more loaded with maintenance work, there is a more intensive consumption of spare parts located in the racks, which must be replenished.

The basic technology for replenishing racks with spare parts has disadvantages: low efficiency of replenishing racks (request for spare parts to the warehouse, transporting them by vehicle to the rack); additional downtime of maintenance wagons while waiting for spare parts.

5 Results

To eliminate the shortcomings in the existing technology, another technology is proposed for replenishing work racks with spare parts (Fig. 2). A working rack is understood as a rack located in the real maintenance area.

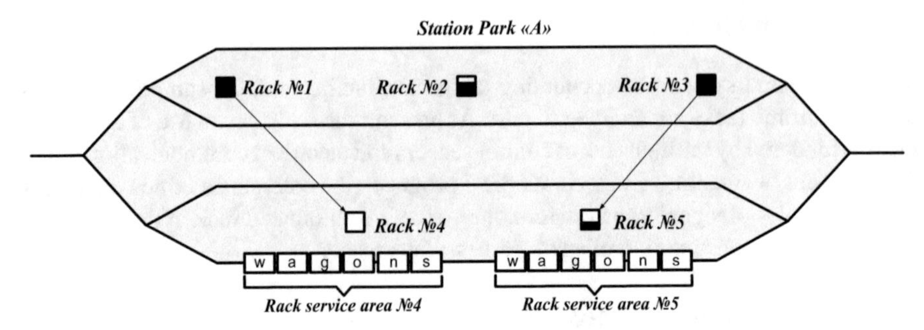

Fig. 2. The proposed technology for replenishing work racks with spare parts.

The essence of the technology lies in the fact that the work racks are replenished not from the warehouse, but from other nearby completed racks that are not involved in maintenance work (at a given time). Replenishment of the source racks is carried out from the warehouse by transporting spare parts on a vehicle.

The proposed technology ensures prompt replenishment and completion of working racks due to the redistribution of spare parts between the racks, which reduces the downtime of wagons waiting for spare parts. Reducing the time for finding the serviceable parts will potentially improve the quality of repairs without cars setting out.

To implement this technology, it is required to equip the fleet (several fleets) of the sorting station with automated racks with the function of accounting of the availability of spare parts.

Automated racks differ in design, overall dimensions, electronic filling of control sensors. At the same time, the principle of operation of all such racks is usually based on tensometry. A schematic diagram of a typical automated rack is shown in Fig. 3.

Removable cells with tensometry sensorswith the required reference value for the same type of parts in conditions of uneven consumption of spare parts in different areas of car maintenance expand the functionality of the device.

The receiving device with storage memory allows to accumulate information about the type and number of storage elements of the rack cells located at the positions of the uncoupled repair of cars.

The operating unit with a rack selection button provides a view of the occupancy of the cells of other racks from the display of each of the racks included in the automated complex, which increases the efficiency of the search for spare parts by the operator located in the immediate vicinity of the rack.

The electromechanical lock of each storage cell provides access to storage elements from one electronic reader, and reduces the risk of shifting parts of different weights from one shelf to another. Thus, the functionality of the device is increased.

The block for analysis of accounting and workload of rack cells automatically collects information about the load of racks and the consumption of parts for the billing period and gives a command to the rack operator (car repair inspector) to add or reduce the number of sections for the same type of parts [12, 13]. This allows to reduce the time spent on searching for the required part in real time, taking into account changes in the structure of the car traffic.

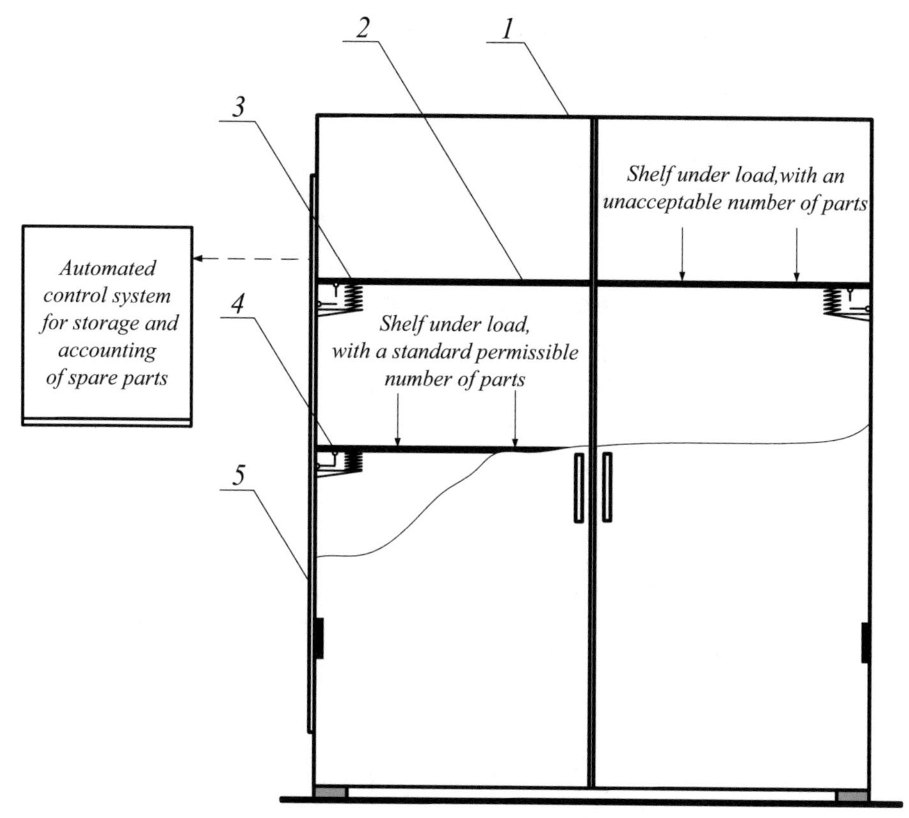

Fig. 3. Schematic diagram of the automated rack: 1 - body; 2 - shelf with storage cells; 3 – sensor; 4 - contactor; 5 - cable box.

6 Findings

The mobile device is a signal receiver with a power source and a display that provides real-time information about the location of the completed nearby racks, the number and type of part. The display of the device has a backlight for working in the dark, and can be equipped with the option of voicing messages (Fig. 4).

Sending messages to a mobile device is carried out automatically via GSM (or LTE) networks via a modem that is integrated into the ACS system for storage and accounting of spare parts.

The device allows the cars repair inspector quickly find out in which of the racks of the fleet there is a part required for repair. The mobile device can only work in an automated system for accounting of the availability and consumption of spare parts [8, 9]. Electronic racks are an integral part of it. Currently, the widespread use of automated racks in fleets of sorting station is hindered by the high cost of control sensors and electrical communications. With the development of scientific and technological progress (the emergence of new materials, other principles of signal transmission), the cost of an

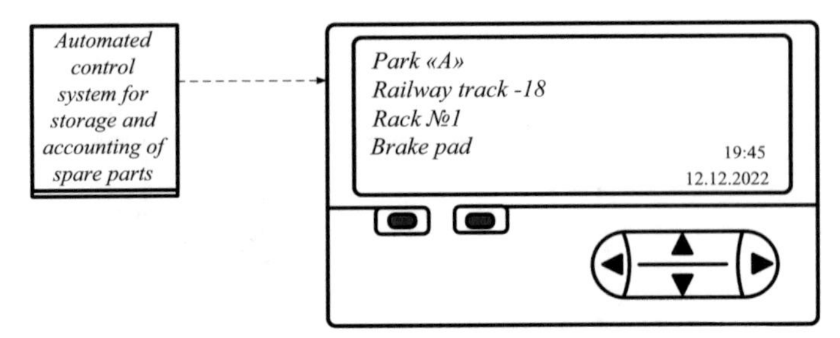

Fig. 4. Mobile wireless device.

automated rack in the system for accounting and consumption of spare parts will decrease [10, 11].

7 Discussion

The automated system for accounting of the availability and consumption of spare parts works as follows. The operator brings the electronic key 6 to the electronic reader 5 installed on the cabinet door 3 (Fig. 5). Numeric key codes that identify the operating personnel who are allowed access to this cabinet are stored in the microcontroller 8.

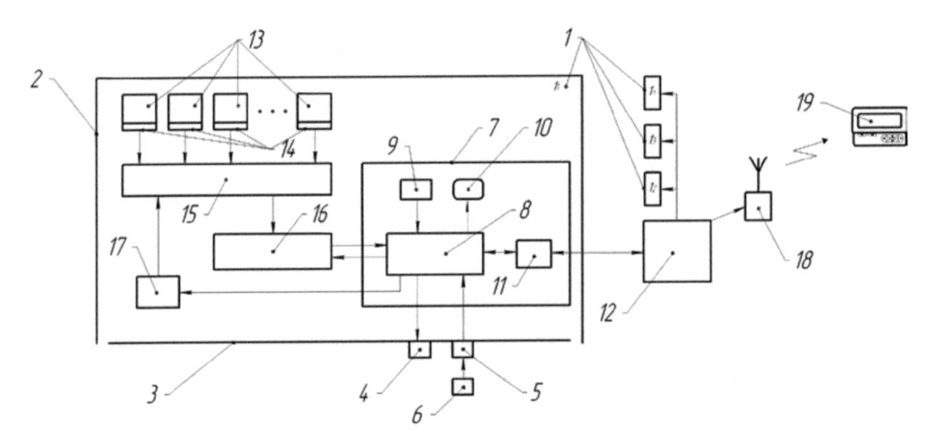

Fig. 5. Automated system for accounting of the availability and consumption of spare parts.

If the key code 6 matches one of the codes stored in the microcontroller 8, the position code of the selected cell is received at the control input of the switching unit 15 from the output of the cell address generation unit 17, the microcontroller gives a command to open the lock 4. After opening the cabinet 2, the operator makes a bookmark or removal of products stored in the cells of the rack 13. When placed on the removable cells of the rack 13 of the same type of storage elements (brake checks, shoes, bolts, sealing rings, shoe suspension rollers, etc.), the state of the cells is determined by tensometry sensors 14, which convert the weight parameter of the parts into a proportional the electrical signal generated by the sensor 14 and is fed to one of the signal inputs of the switching unit 15, the output of which is connected to the input of the analog-to-digital converter 16. From the output of the analog-to-digital converter 16, the multi-bit digital signal code of the selected cell enters the microcontroller 8, which generates a digital data packet that contains the information about the cell address and its current state.

Information about the number and type of parts in manual mode is displayed on the display 10 of the input device 7 at the request of the rack operator from the keyboard for selecting the state of cells 9. Data on other racks is retrieved from the receiving device with storage memory using a microcontroller.

If necessary, the car repair inspector, being in close proximity to the rack, using the operating unit with the rack selection button, input device 7, can view information on the occupancy of the cells of other racks included in the automated system on the display 10. After receiving information about the state of occupancy of the rack cell, the car repair inspector opens the electromechanical lock 4 of the requested rack cell with one electronic reader. After removing the required part from the rack cell, the storage section is closed.

Automatically, information from racks 1 enters a remote terminal for collecting, processing and transmitting information 12 through a data transmission device 11 via wired or radio channels, where information for the billing period about the most loaded racks and their cells is evaluated in the block for analyzing the accounting and workload of rack cells. Using this information, the number of cells of the most loaded racks can be adjusted, in order to do it, the cells are designed as a removable structure.

From the remote terminal 12, information about the occupancy of the cells is transmitted by feedback to all racks of the automated complex 1. In addition, information about the quantity and type of part is transmitted to the mobile receiver 19 of car repair inspector from the remote terminal 12 through the communication unit 18 in the form of short messages in automatic mode via GSM (or LTE) networks. This allows the rack operator to improve the search for complete racks in operational conditions while being in different part of the fleet.

8 Conclusion

Thus, the developed automated system for accounting of the availability and consumption of spare parts at railway transport facilities will provide:

- Improvement of the information content of accounting of the availability and consumption of spare parts in the racks located at the positions of the cars repair without setting out;
- Accumulation of information about the number and type of storage elements by each rack included in the automated complex;
- Operational search for completed racks and storage cells;
- The ability to change the number of storage cells of the same type of parts, depending on the load on the racks.

The automated system for accounting of the availability and consumption of spare parts can be used in railway transport to control the accounting of the availability and consumption of spare parts in fleet racks located at the positions of cars repair without setting out. The technical result is improvement of the efficiency and information content of automated accounting of the availability and consumption of spare parts of the racks located at the positions of cars repair without setting out.

References

1. Electronic storage racks and automatic inventory of spare parts. http://www.tdisie.nsc.ru/Rus/stellag_rus.html. Accessed 12 Oct 2022
2. Kiselev, G.G., Klyukanov, A.V.: Automated complex of racks for storage and accounting of spare parts. Patent for Invention RU 2778490 C2, 22.08.2022, Application No. 2019125420 dated 06.08.2019
3. Rules for labour protection during maintenance and repair of freight cars. No. 57 approved on 17 Jan 2013, 40 p (2013)
4. Klyukanov, A.V., Parenyuk, M.A.: Technology of replenishment of racks with spare parts of cars using a mobile wireless device. Volga Reg. Transp. Periodical **4**(64), 30–34 (2017)
5. Vasilev, V.L., Gapsalamov, A.R., Akhmetshin, E.M., Bochkareva, T.N., Yumashev, A.V., Anisimova, T.I.: Digitalization peculiarities of organizations: a case study. Entrep. Sustain. **7**, 3173–3190 (2020)
6. Gubernatorov, A.M., Chistyakov, M.S.: Convergence of digital technologies and industrial potential of manufacturing industries in the formation of the platform approach "Industry 4.0". In: Management of the Economy: Methods, Models, Technologies: Materials of the XX International Scientific Conference, pp. 62–65. Ufa: Ufa State Aviation Technical University (2020)
7. Koshekov, K., Kobenko, V., Koshekov, A., Moldakhmetov, S.: Hand-written character structure recognition technology on the basis of identification measurements. ARPN J. Engi. Appl. Sci. **15**(21), 2555–2562 (2020)
8. Kalantayevskaya, N., Koshekov, K., Latypov, S., Savostin, A., Kunelbayev, M.: Design of decision-making support system in power grid dispatch control based on the forecasting of energy consumption. Cogent Eng. **9**(1), 2026554 (2022)
9. Koshekov, K.T., et al.: Modernization of vibration analyzers based on identification measurements. Russ. J. Nondestr. Test. **54**(5), 328–334 (2018). https://doi.org/10.1134/S106183091805008X
10. Nadirov, K.S., et al.: The study of the gossypol resin impact on adhesive properties of the intermediate layer of the pipeline three-layer rust protection coating. Int. J. Adhes. Adhes. **78**, 195–199 (2017)
11. Kolesnikov, A.S., et al.: Processing of non-ferrous metallurgy waste slag for its complex recovery as a secondary mineral raw material. Refract. Ind. Ceram. **62**(4), 375–380 (2021). https://doi.org/10.1007/s11148-021-00611-7
12. Nadirov, K.S., et al.: Examination of optimal parameters of oxy-ethylation of fatty acids with a view to obtaining demulsifiers for deliquefaction in the system of skimming and treatment of oil: a method to obtain demulsifier from fatty acids. Chem. Today **34**(1), 72–77 (2016)
13. Kolesnikova, O., et al.: Thermodynamic simulation of environmental and population protection by utilization of technogenic tailings of enrichment. Materials **15**, 6980 (2022)

Application of Immersion Technology in the Creation of Digital Prototypes of Green Energy Objects

Yu. Protasova[2], A. Rodina[1], A. Borisenko[1], V. Nemtinov[1](✉), and P. K. Praveen[3]

[1] Department of Computer-Integrated Systems in Mechanical Engineering, Tambov State Technical University, 106, Sovetskaya Street, Tambov 392000, Russian Federation
nemtinov.va@yandex.ru

[2] Department of Management, Service and Tourism, Tambov State University Named After G R Derzhavin, 33, Internatsionalnaya Street, Tambov 392000, Russian Federation

[3] AGG Lifesciences and Safety Solutions LLP, 1102, Lodha Supremus, Sakivihar Road, Opp. Sakivihar MTNL, Powai, Mumbai 400072, Maharashtra, India

Abstract. In this article, the issues of developing a technology for creating digital prototypes of green energy facilities and their use in the educational process for training of specialists in the energy industry during lectures, practical classes and exams are considered. An immersive educational environment was created using Twinmotion, Bigscreen and 3D Vista Virtual Tour Pro software. A digital prototype of an energy complex was created in the Twinmotion software environment, including the main sources of alternative electric energy: wind turbines and solar panels, providing immersive architectural 3D visualization. The created immersive environment is used for the following educational purposes: mastering the actions of specialists during emergency situations; teaching students; taking educational quests by schoolchildren.

Keywords: Virtual Environment · Virtual Modeling · Digital Transformation · Virtual Tour · Twinmotion · E-learning · 3D Vista Virtual Tour pro

1 Introduction

Since its start, the energy industry has experienced rapid development. As the result of this process, which began with the use of steam power, enterprises have now witnessed the consequences of the Fourth Industrial Revolution with rapid technological development. The widespread use of digital technologies and their continued development has changed business models, ways of doing business and making decisions in many areas of manufacturing and service, including energy industry as well.

In the modern world, information technologies are being increasingly introduced to all branches of industrial production. In accordance with the global trend towards decarbonization, in order to achieve climate stabilization, in particular, zeronet greenhouse gas emissions by 2050–2070, the transformation of global energy is also taking place.

© The Author(s), under exclusive license to Springer Nature Switzerland AG 2023
A. Gibadullin (Ed.): DITEM 2022, LNNS 683, pp. 141–151, 2023.
https://doi.org/10.1007/978-3-031-30926-7_14

And at the same time, new technologies are becoming a significant factor contributing to the changes in the electric power industry: big data processing, machine learning, cloud computing, augmented and virtual reality, etc.

The relevance of transformation is due to the fact that digitalization as a process is one of the priorities in the country's strategic development and is crucial to stable economic growth. Digitalization is of great importance for the energy security of regions and also for solving environmental problems; it helps to identify the reasons for the need to transform the electric power industry and proposes a set of measures aimed at introducing digital technologies and developing new business models, services and markets based on digital solutions that ensure the development of innovative technologies and services in the field of energetics. In the article [1], the authors present the factors contributing to development of the energy industry, as well as competitive trends in the market of goods and services. Within the framework of digital transformation, strategic objectives for the development of the electric power industry are defined and the authors propose a model of digital transformation of the electric power industry, which contributes to improving the reliability of power systems by reducing costs of production and introducing new technologies based on digital platforms.

In the context of climate change, the energy crisis caused by the COVID-19 pandemic and the embargo on the supply of raw materials from Russia, great expectations are placed on the development of renewable energy sources in terms of meeting energy needs. However, renewable energy sources also have several disadvantages. In the most dynamically developing sectors of solar and wind energy, the main problems include the storage of this energy and ensuring security of supply. The solution to these problems is possible due to digital transformation of renewable energy production together with the market entry of new players who are implementing digital business models in the field of renewable energy [1].

The article [2] analyzes modern digital technologies in the energy sector on the example of digital substations. The basic principles and advantages of digital technologies are presented, as well as generally accepted principles are formulated. It is shown that intelligent data processing is necessary for digitalization of the energy industry. According to the authors, digital transformation is done through the introduction of modern technologies into business processes of an enterprise, which implies fundamental changes in management approaches, corporate culture and external communications. Digitalization, Industry 4.0 and the Internet of Things are the components of digital transformation in the energy sector [3, 4].

The first steps towards digital transformation of production systems have been taken since the mid-90s, but notable progress in this direction occurred only with the advent and development of technologies such as the Industrial Internet of Things, Big Data processing and Cognitive Computing, Virtual Reality (VR) and Augmented Reality (AR) [5–7]. On the basis of augmented and virtual reality (AR/VR) technologies, it is possible to create various expert systems, interactive electronic technical manuals, display information about the operating modes of equipment (including telemetry), etc. As a result of introduction of AR/VR technologies, labor productivity increases due to optimization of personnel movements, reduction of production time; as well as the effectiveness of training of service personnel.

In this regard, this paper considers the issues of digital transformation of green energy objects in the educational process.

2 Materials and Methods

When implementing the technology of creating a digital prototype of an energy complex as part of a production cluster, at the first stage, a layout of the entire territory is built, including the territories closest to the plant with the natural ecosystem around it using the Twinmotion software, which provides architectural and landscape 3D visualization [8–10].

Such a tool as Twinmotion Presenter should also be mentioned, as it allows to develop the project in the form of a separate file with all the necessary resources for offline viewing. Both the Twinmotion system itself and the Twinmotion Presenter support virtual reality glasses, in particular Oculus Quest, HTC VIVE [11].

A 3D scene is created in the Twinmotion system (see Fig. 1), including terrain, roads, municipal infrastructure systems, as well as imported three-dimensional models of industrial facilities made in 3D modeling programs (SketchUp, Blender, Archicad, etc.) [12]. Further, the program implements high-quality rendering of images and videos necessary for the development of a virtual tour (including 360° format), as well as spherical panoramas.

Fig. 1. Visualization of the digital model of the energy complex (general view).

At the next stage, an interactive virtual tour is created using the 3D Vista Virtual Tour Pro software environment, which uses multimedia materials obtained at the previous stage [8, 13–15]. A virtual tour of the production cluster, which includes the energy complex, is located at: https://heritage.tstu.ru/memorial/directaccess/zavod/index.htm.

This virtual tour can be used for various educational purposes, such as staff training, mastering the actions during emergency situations, performing educational quests on the territory of the complex, etc. For communication inside the virtual space, the Live

Guide Tours tool, which a part of 3D Vista Virtual Tour Pro package,is used for video conferencing in real time. The Live Guide Tours system does not support special virtual reality devices such as glasses and helmets and it is operated on computers, laptops, tablets and mobile devices [15, 16]. Verification of the knowledge obtained in the framework of e–learning is carried out through quests using hot spots (the so-called "treasure hunt"), question cards (Quiz Cards) and quizzes, the results of which can be transferred to the LMS Moodle system [17–24].

The developed educational content, in particular lecture material and the database of questions for testing the acquired knowledge in the questform are presented below.

3 Results

Wind Turbines. Windmills have been helping humanity to convert wind energy into many other useful forms of energy over the past 20 centuries. Modern industrial wind turbines are capable of converting wind energy into electricity with high efficiency. This is achieved due to the optimal shape of the blades, developed using modern means of aerodynamic analysis, as well as other special devices that increase the efficiency of wind generators (see Fig. 2). The aerodynamic profile of the blades allows the rotor of the wind turbine to spin. The blades are positioned at a certain angle of attack relative to the incoming air flow. The linear velocity of the blade relative to the air medium increases in the direction from the root of the blade to its tip. Therefore, the blades have a geometric torsion along the entire length, which gives a more uniform distribution of aerodynamic forces. The speed of rotation is low enough to prevent its physical destruction due to centrifugal forces, as well as to reduce the noise level. However, the shaft of the electric generator requires a high rotation speed, so the rotation from the rotor is transmitted to the generator shaft through step-up gearboxes. Also, a special braking mechanism is placed in the nacelle of the wind turbine to slow down the rotation of the blades in excessive wind.

The generated electrical energy is transmitted via cables to the base of the wind generator, where the step-up transformer is located. To achieve maximum efficiency of energy generation, the axis of rotation of the rotor should be parallel to the wind flow. However, the wind direction may change at any time. To measure the direction and speed of the wind, an anemometer is located on top of the nacelle (see Fig. 3). When the wind direction changes, the signal from the anemometer enters the electronic controller, which in turn sends the corresponding control signal to the rotary mechanisms that corrects the position of the wind generator relative to the incoming wind flow. Thus, the wind turbine will always be aligned in the direction of the wind.

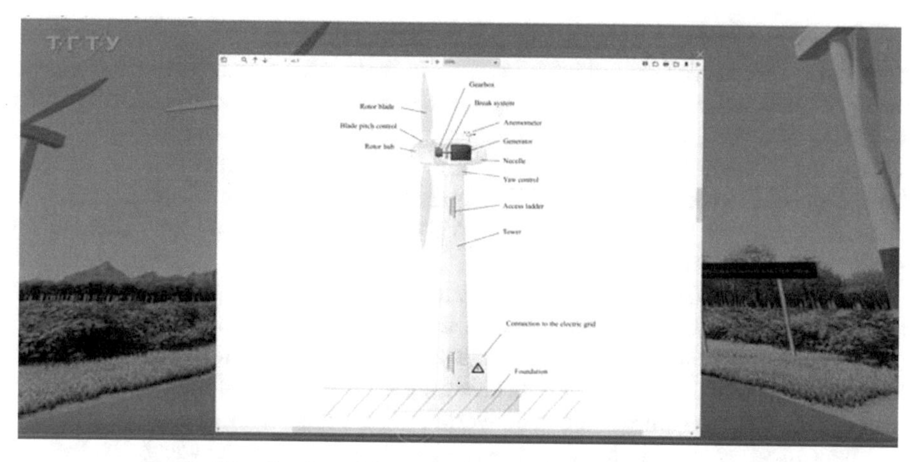

Fig. 2. Visualization of the general view of the wind turbine design.

Fig. 3. Visualization of the main nodes of the wind generator.

In addition, depending on the wind speed, the angle of attack of the blades is also adjusted by a special rotary mechanism (yaw control) of the blades. As a result, the blades are always at an optimal angle to the incoming wind flow.

Due to the fact that the blades take part of the kinetic energy of the wind, the speed of the wind flow behind the turbine is less than in front of it. To convert 100% of kinetic energy of the wind into mechanical energy, it is necessary that the wind speed behind the turbine is zero (the flow should be completely stopped), which is physically impossible. This means that there is a theoretical marginal efficiency that a wind turbine can achieve. This limit is known as the Betz limit (law), which states that a wind generator can take no more than 59.3% of the air flow power, while the flow slows down three times when passing through the rotor.

To test the knowledge gained in the course of theoretical and practical studies, the authors have created a database of questions, the answers to which can be obtained during the quests. Table 1 shows a fragment of the database of test questions for assessing the knowledge about wind turbines (correct answers are marked in bold).

Solar Panels. The entire vegetation world captures the energy of sunlight and uses it to convert water and carbon dioxide into simple sugar – glucose, which is a fuel for plants obtained with the help of the sun. Not only plants can convert sunlight into energy. The rays of the sun can be converted into electric current using solar batteries.

Table 1. A fragment of the database of test questions about wind turbines in the form of a quest.

Test Questions	Response options	The number of points for the correct answer	Maximum response time (minute), ∞ - no restrictions
At what wind speed do wind generators give nominal power?	**10 m/s**	5	3
	2 m/s		
	20 m/s		
	5 m/s		
When the wind direction changes … sends a signal to the rotary mechanisms to adjust the position of the turbine	**Anemometer**	10	5
	Generator		
	Rotary mechanisms		
	Transformer		
How many blades should there be in a wind turbine to achieve the greatest degree of efficiency?	One	10	3
	Two		
	Three		

The basis of the solar battery is made up of pure silicon crystals. In nature, silicon is found only in the form of sand, from which plates with a thickness of 180 microns are made. A small amount of boron and phosphorus is applied to the silicon wafer. Free electrons appear in the silicon layer with the addition of phosphorus, and missing electrons, so-called "holes", appear in the silicon layer with boron additives. When a quantum of light hits a solar battery, particles begin to move from one layer to another, so that an electric current arises. In the case of directed sunlight, electricity is collected at each point of the silicon wafer. Figure 4 shows the visualization of the main nodes of the solar panel.

If one small plate can be enough to operate a small pocket flashlight, then when they are connected, the battery power increases significantly. Even if it's cloudy outside, the batteries still receive energy. During the day, batteries accumulate electricity, and at night they give it away. The service life of solar panels is about 25 years.

Fig. 4. Visualization of solar panels (in lecture mode with a "live guide").

Photovoltaic panels (solar panels, solar batteries) can be installed, for example, on the roofs of industrial and residential buildings, such as warehouses, industrial workshops, shopping and entertainment centers, sheds, etc.

To increase the efficiency of electric energy generation, charging stations for electric vehicles are widely used, which in addition to solar panels contain additional buffer batteries (Photovoltaic/Battery Energy Storage/Electric Vehicle Charging Stations - PBES). In particular, the examples of operating solar power plants used to charge electric vehicles are Evergreen Solar Fuel Station in Frankfurt (generates 21 kWh), Beautiful Earth Group solar EV Charger in New York (generates up to 5.6 kWh), E-Move Charging Station (generates up to 2 kWh), Solar Grove in San Diego (generates up to 14.72 kWh) [25].

In Russia, an example is a charging station in St. Petersburg (generates up to 2.4 kWh) [25]. Electric vehicles connected to the grid, as well as a buffer battery, serve as an energy storage and reduce uneven energy generation by solar panels due to uneven lighting.

Figure 5 shows visualization of a fragment of a digital model of an energy complex, including a charging station for electric vehicles [26]. Electric vehicles are charged from photovoltaic modules installed on a special canopy.

Table 2 shows a fragment of the database of test questions for assessing the knowledge about solar panels (correct answers are marked in bold).

Fig. 5. Visualization of a fragment of a digital model of an energy complex, including an electric vehicle charging station.

Table 2. A fragment of the database of questions for assessing the knowledge about solar panels in the form of a quest.

Test Questions	Response options	The number of points for the correct answer	Maximum response time (minute), ∞ - no restrictions
Is there enough sunlight to produce electrical current?	No, sunlight cannot be converted into electric current	5	3
	Yes, it is possible to use sunlight as a source for generating energy		
What kind of sun rays are collected by solar panels?	Only direct	10	5
	Only indirect		
	Direct and indirect		
Under what condition is rain able to clean solar panels fromdirt?	Rain is not able to clean solar panels	10	5

(continued)

Table 2. (*continued*)

Test Questions	Response options	The number of points for the correct answer	Maximum response time (minute), ∞ - no restrictions
	When the panel is tilted more than 20°		
	When the slope is less than 20°		
What devices can be connected to a solar panel system?	Batteries only	5	3
	Appliances		
	Car		
	Everything that runs on electricity		
Which system produces direct current from solar energy?	**Photovoltaic system**	5	3
	Thermal system		
During the ….. Time diodes do not allow current to flow in the opposite direction from the battery into the panel and by heating it to lose power reserve	**Night**	5	3
	Rainy		
	Sunny		
Do silicon crystals in solar panelswear out?	**Yes, over time, silicon crystals lose their productive capacity**	5	3
	No, silicon crystals do not need to be replaced		

4 Discussion

In order to confirm the feasibility of digital representation of green energy facilities, a study was conducted to assess obtained knowledge about them during the training of specialists in the energy industry. The test results of two groups of students with 14 people in each group, conducted with immersion in the virtual space of the energy complex as part of the production cluster (group 1) and without it (group 2), showed a 32% higher proportion of correct answers to the questions of the proposed test for the first group. This is another confirmation of the expediency of using immersive educational environment to improve the effectiveness of learning. This ensures the achievement of the planned learning outcomes correlated with the indicators of competences acquisition.

5 Conclusion

As a result of the conducted research, the authors proposed the technology for creating a digital prototype of green energy facilities and their application in the educational process when training specialists in the energy industry during lectures, practical and exams. Digital representation of green energy facilities in the educational process creates a wide range of various forms of educational communication based on the capabilities of modern learning tools and modern information and communication technologies, allows providing conditions for improving the quality of education, exchange of opinions, mutual consultation.

References

1. Pakulska, T., Poniatowska-Jaksch, M.: Digitalization in the renewable energy sector-new market players. Energies **15**(13), 4714 (2020)
2. Turovets, J., Proskuryakova, L., Starodubtseva, A., Bianco, V.: Green digitalization in the electric power industry. Foresight STI Gov. **15**(3), 35–51 (2021)
3. Trunova, L.G.: Digital transformation of the electric power industry. In: Paper Presented at the AIP Conference Proceedings, vol. 2434 (2022)
4. On the digital transformation of the energy industry. https://energypolicy.ru/o-czifrovoj-transformaczii-energeticheskoj-otrasli/neft/2021/19/05/. Accessed 15 Oct 2022
5. Guidi, G., Russo, M., Angheleddu, D.: 3D survey and virtual reconstruction of archeological sites. Digit. Appl. Archaeol. Cult. Herit. **1**(2), 55–69 (2014)
6. Nisiotis, L., Alboul, L., Beer, M.: Virtual museums as a new type of cyber-physical-social system. In: De Paolis, L.T., Bourdot, P. (eds.) AVR 2019. LNCS, vol. 11614, pp. 256–263. Springer, Cham (2019). https://doi.org/10.1007/978-3-030-25999-0_22
7. Nemtinov, V.A., Gorelov, A.A., Nemtinova, Y.V., Borisenko, A.B.: Visualization of a virtual space and time model of an urban development territory. Sci. Vis. **8**(1), 120–132 (2016)
8. Nemtinov, V.A., Gorelov, A.A., Nemtinova, Y.V., Borisenko, A.B.: Implementation of technology for creating virtual spatial temporal models of urban development history. Sci. Vis. **10**(3), 99–107 (2018)
9. Nemtinov, V., Egorov, S., Borisenko, A., Morozov, V., Nemtinova, Y.: Support of design decision-making process using virtual modeling technology. In: Gibadullin, A. (ed.) DITEM 2021. LNNS, vol. 432, pp. 70–77. Springer, Cham (2022). https://doi.org/10.1007/978-3-030-97730-6_7
10. Egorov, S.Y., Sharonin, K.A.: Automated decision making in the problem solving of objects layout for chemical and refining industries using expert software systems. Chem. Pet. Eng. **53**(5–6), 396–401 (2017)
11. Nemtinov, V., Bolshakov, N., Nemtinova, Y.: Automation of the early stages of plating lines design. In: MATEC Web Conference, vol. 129, p. 01012 (2017)
12. Epic, Games: a cutting edge real-time architectural visualization tool – Twinmotion. https://www.twinmotion.com/. Accessed 15 Oct 2022
13. Pozdneev, B., Tolok, A., Ovchinnikov, P., Kupriyanenko, I., Levchenko, A., Sharovatov, V.: Digital transformation of learning processes and the development of competencies in the virtual machine-building enterprise environment. J. Phys.: Conf. Ser. **1278**, 012008 (2019)

14. Karpushkin, S.V., Krasnyanskiy, M.N., Malygin, E.N., Mokrozub, V.G.: Determination of maximum productivity the technological system of multi-product chemical plant. In: IOP Conference Series: Materials Science and Engineering, vol. 971, p. 032056 (2020)
15. Duda, J., Oleszek, S.: Concept of PLM application integration with VR and AR techniques. In: Lalic, B., Majstorovic, V., Marjanovic, U., von Cieminski, G., Romero, D. (eds.) APMS 2020. IAICT, vol. 592, pp. 91–99. Springer, Cham (2020). https://doi.org/10.1007/978-3-030-57997-5_11
16. 3DVista. 3DVista – Virtual Tours, 360° video and VR software.https://www.3dvista.com/. Accessed 16 Oct 2022
17. 3DVista. Virtual Tours in E-Learning, Training & Quizzing. https://www.3dvista.com/en/blog/virtual-tours-in-e-learning-training-quizzing-v2/. Accessed 15 Oct 2022
18. Pereira, G.: and. In: Schweiger, G. (ed.) Poverty, Inequality and the Critical Theory of Recognition. PP, vol. 3, pp. 83–106. Springer, Cham (2020). https://doi.org/10.1007/978-3-030-45795-2_4
19. Unal, A., Karakuş, M.A.: Interacting science through Web Quests. Univers. J. Educ. Res. **4**(7), 1595–1600 (2016)
20. Nemtinov, V., Zazulya, A., Kapustin, V., Nemtinova, Y.: Analysis of decision-making options in complex technical system design. J. Phys. Conf. Ser. **1278**(1), 012018 (2019)
21. Borisenko, A.B., Nemtinov, V.A.: The task of integrating photovoltaic panels into the infrastructure of charging stations. In: Computer integration of production and IPI-technologies, Collection of materials of the X All-Russian Conference. Orenburg, pp. 275–279 (2021)
22. Krol, O., Sokolov, V.: Development of models and research into tooling for machining centers. J. Eastern-Eur. J. Enterp. Technol. **3**(12) (2018)
23. Alekseev, V., Lakomov, D., Shishkin, A., Maamari, G.A., Nasraoui, M.: Simulation images of external objects in a virtual simulator for training human-machine systems operators. J. Phys.: Conf. Ser. **1278**, 012032 (2019)
24. Onyesolu, M.O., Eze, F.U.: Understanding virtual reality technology: advances and applications. In: Advances in Computer Science and Engineering, pp. 53–70. Tech Open Access Publisher, Rijeka (2011)
25. Arhun, S.: Projects and models of solar charging stations for electric cars. Bull. Kharkov Natl. Automob. Highw. Univ. **80**, 45 (2018)
26. Laikov, D.A.: Autonomous gas stations for electric vehicles. New direction in architecture and power engineering. Actual problems of architecture and urban planning: materials of the Republican Student Scientific and Technical Conference, 75th Student Scientific and Technical conference of BNTU, pp. 92–97 (2019)

Algorithm Creating for Extracting Feature Lines to Control the Quality of Industrial Products

Oxana Kozhukalova[1], Tatiana Efremova[2], Michail Makovetskij[3], Natalya Gavrikova[4](✉), and Marat Arifullin[5]

[1] Federal State Budgetary Educational Institution of Higher Education, "Moscow Pedagogical State University", Moscow, Russia
[2] National Research Mordovia State University, Saransk, Russia
[3] Moscow Witte University, Moscow, Russia
[4] GAOU VO MGPU, "Moskovskiy Gorodskoy Pedagogicheskiy Universitet", Moscow, Russia
zzznataljazzz@rambler.ru
[5] State University of Management, Moscow, Russia

Abstract. The article reveals the processes features of algorithm creating for extracting feature lines for the quality control of industrial products. Algorithms for extracting characteristic lines of the industrial parts as image contours are considered, an algorithm for determining the characteristic lines of the external images contour for industrial parts is proposed and analyzed, and an example of a software implementation of the proposed algorithm and results are given. Based on the proposed algorithms for determining the characteristic lines of the outer contour of industrial parts' images, a software system was developed in the Delphi 6.0 programming language. The program extracts feature lines, external image contours of industrial parts and conducts a number of studies, including the shape of industrial parts. Abstract should summarize the contents of the paper in short terms, i.e. To further develop and increase the efficiency of the process of passing the characteristic lines of the contours of images of industrial parts, it is necessary to develop criteria for the correct shutdown of algorithms, as well as ways to process the protrusions of images (thickness of 1 pixel) more clearly.

Keywords: Algorithm · Feature Lines · Recognition · Image · Control · Quality · Industrial Parts

1 Introduction

One of the current directions in the development of modern applied science is the widespread introduction of computer tools in various practical scientific fields and industry [7]. In particular, computer technologies with a set of specialized software tools are used to diagnose graphic objects, quickly process information and exchange data, as well as to train future specialists, etc. [1, 8, 10].

At the present stage, the advanced information technologies, new methods and algorithms for processing, analyzing and synthesizing images in various scientific fields and industry has led to the creation of new areas, for example, telemedicine, which provides

© The Author(s), under exclusive license to Springer Nature Switzerland AG 2023
A. Gibadullin (Ed.): DITEM 2022, LNNS 683, pp. 152–159, 2023.
https://doi.org/10.1007/978-3-031-30926-7_15

for the quality diagnostics of industrial products [5]. So, product quality management in industrial sectors is one of the primary tasks.

As researchers consider [8, 10], each type of industrial product has its own quality signs: the appropriate geometric shape, dimensional accuracy and characteristic surface. Highlighting these features in automatic mode is one of the main tasks that must be solved by algorithm creators and software. This problem is most relevant for industrial enterprises with large production volumes and responsible purpose of products. Increasing the reliability of information about the industrial products' quality contributes to the modern computer technology and the possibility of improving algorithms for automated image recognition and analysis. Thus, the research object is the analysis of algorithms for the allocation of characteristic lines for managing the industrial products quality.

The research goal is to study the algorithm for allocating characteristic lines for quality management of industrial products.

2 Introduction

In modern time, many businesses and especially national entire industries are under increasing competitive pressure. The survive guarantee and long-term success of the organization directly depends on product and service quality. This led to the allocation of an important place in the overall management system of the enterprise.

It is to ensure the high competitiveness of enterprises that the quality of their products is of great importance. Recent world-leading manufacturers achieve high quality products solely on the basis of the effective quality management system. The main problem, for example, in Russian-China trade relations lies in the fact that these countries conclude agreements for the purchase of products with Russian enterprises. These documents have implemented and operate a quality management system as the basis of modern enterprise management. However, the crisis phenomena in the Russian economy in recent years have led to the fact that in modern enterprises the real implementation and functioning of the quality management system is almost rare. And, accordingly, there is no corresponding methodology and practical experience of its creation [4]. In the available scientific publications, for example [7, 8, 10], there are no practically justified effective methods for analyzing the causes and eliminating inconsistencies in the quality management of products of industrial enterprises.

In developing the structure of the quality management system, industrial companies must adhere to the following stages of its construction:

- Identify the processes required for the quality management system and their application throughout the organization.
- Determine the sequence and interaction of these processes
- Define the criteria and methods necessary to ensure the effectiveness of both the implementation and management of these processes.
- Ensure that the resources and information necessary to maintain these processes and their constant monitoring are available.
- Carry out continuous monitoring, measurement, where possible, and analysis of these processes.

- Apply the measures necessary to achieve the planned results and continuous improvement of these processes.

Production of industrial parts and products for various purposes, manufacture and installation of metal structures, shaped parts and pipeline blocks necessarily include visual and measuring quality control at all stages.

Visual-measuring quality control of industrial components and products is the basic method of non-destructive testing and is always used before other control methods.

It is important to carry out visual quality control of industrial products at all stages of production or assembly. If we apply visual and measuring quality control of industrial component and products only at the last stage, there is a risk of missing defects that have arisen during the production process. Often during the inspection, specialists discover defects, correction of which in the finished object can take considerable time. So, neglecting quality control can lead to serious breakdowns or accidents.

When performing visual and measuring quality control of products, materials, semi-finished products, parts, blanks, determine by the following:

- Absence of surface defects: cracks, detachments, ruptures, shells, captivity, hair, faces and other defects in the manufacture of materials, as well as damage caused during transportation and assembly of parts – corrosion, deformation, etc.;
- Compliance of the dimensions of products, materials, semi-finished products, parts, blanks with the initial requirements;
- Correspondence of the geometric shape of the components' edges;
- Absence (or presence and dimensions) semi-finished products containing welds, surface defects in welding (surfacing): cracks, pores, fistulas, undercuts, burns;
- Compliance with the requirements for the dimensions of the weld (surfacing) and the detected defects.

Without going into the details of the technology for the preparation of samples for quality assessment in the production of industrial products, it should be noted that scanned and digitized images of sulfur prints from transverse templets are used as input data.

One of the important stages of image analysis is the correct and fast selection and passage of the image contour. Let's analyze the known algorithms for determining the external contour of the image. The advantages of the "Square Tracing" algorithm [3] are the simplicity of implementation and speed. The disadvantages are the ineffective study of images with diagonal sides and images that have a branch thickness of 1 pixel. This algorithm, despite its simplicity and speed, cannot be effectively used to determine the complex shape contour.

An improvement of the "Square Tracing" algorithm is the "Moore's Neighbor tracing" algorithm, which, unlike the previous one, is more efficient when passing images with arbitrary sides. Greater efficiency is achieved by increasing the number of neighboring points that pass the verification process. The advantages of this algorithm are an increase in processing accuracy and the ability to process more complex images. Nevertheless, disadvantages are an increase in the number of neighboring points passing the verification process; an increase in computational complexity and reduced performance, as well as the possibility of incorrect termination of the algorithm.

The "Radial Sweep" contour algorithm is based on the "Moore's Neighbor tracing" algorithm, but it has the improvement that each subsequent point of the circuit is tied to the previous one. In addition, another criterion for stopping the algorithm has been added (in case of hitting a point that already belongs to the path). This algorithm advantages are an additional stop criterion, according. So, it is more effective to avoid the algorithm looping and correctly complete its work, as well as the ability to work with a complex contour lines. The disadvantages are an increase in computational complexity, as well as not completely correct algorithm completion. Another algorithm for passing the image contour is the algorithm "The Pavlidi's algorithm" [5]. This algorithm advantages are the high quality of work with images that do not have a complex contour line, the ability to use any point of the contour as a starting point as well as a high working speed. The disadvantages are the complexity of implementation, problems with processing images with traces of the contour (thickness of 1 pixel), and also imperfect algorithm completion criteria.

Of great importance for the of the algorithm efficiency and operability for passing along the characteristic lines of the image contour are the criteria for algorithm termination. Among the known termination criteria are the following [8].

The algorithm has visited the starting point n-fold. This criterion is effective with a predetermined amount of algorithm return to the starting point.

The termination criterion after the algorithm visits the starting point twice (Jacobs' stop criterion). This criterion is a special case of the previous one. The advantage is high efficiency when working with simple contours of figures and ease of implementation. But the disadvantage is the choice of the starting point.

The termination criterion when hitting a point that has already been recognized as a path point. The advantage of this criterion is the arbitrary choice of the starting point, the disadvantages are the need to mark the passed-contour points, the inefficiency of determining the protrusions on the path (thickness of 1 pixel).

Let's describe the actual algorithm for highlighting the characteristic lines of the contours of the industrial parts' images.

To determine the characteristic lines of the outer boundary of the images of industrial parts, a method is used by which areas of the image are not analyzed if their color is identical or close to the color of the image with some error Δ [4]. This approach determines the background pixels. The color of the point (pixel) of the characteristic lines is transmitted in the RGB space. To recognize the point of characteristic lines as part of the background, the value of its RGB component must satisfy the following conditions (1–3):

$$\Delta R \geq |R_{fon} - R| \tag{1}$$

$$\Delta G \geq |G_{fon} - G| \tag{2}$$

$$\Delta B \geq |B_{fon} - B| \tag{3}$$

where ΔR, ΔG, ΔB is the value of the possible deviation of RGB components from the specified background values;

R_{fon}, G_{fon}, B_{fon} is background color value in basis RGB;

R, G, B is pixel color value of characteristic lines in the RGB basis;

To improve the results of the algorithm, it is proposed to use a dynamic value of the background color. Dynamic background values are determined by the following rules (4–6):

$$R_{fon} = \begin{cases} \left|\frac{R_{fon}+R}{2}\right| & ; R_{fon} \neq R, \Delta R \geq \left|R_{fon} - R\right|, \\ R_{fon}; R_{fon} = R \end{cases} \tag{4}$$

$$G_{fon} = \begin{cases} \left|\frac{G_{fon}+G}{2}\right| & ; G_{fon} \neq G, \Delta G \geq \left|G_{fon} - G\right|, \\ G_{fon}; G_{fon} = G \end{cases} \tag{5}$$

$$B_{fon} = \begin{cases} \left|\frac{B_{fon}+B}{2}\right| & ; B_{fon} \neq B, \Delta B \geq \left|B_{fon} - B\right|, \\ B_{fon}; B_{fon} = B \end{cases} \tag{6}$$

where ΔR, ΔG, ΔB is the value of the possible deviation of RGB components from the specified background values;

R_{fon}, G_{fon}, B_{fon} is background color value in basis.

RGB; R, G, B is the pixel color value of the characteristic lines in the RGB basis.

Using a dynamic background value reduces the impact of noise and distortion caused by camera sensors. This approach disadvantage is a decrease in the efficiency of work with insufficient image quality (noise and distortion overlap the images of industrial details). So, before starting the program, if necessary, the image requires additional processing. To cut off uninformative points, as well as to overcome some of the shortcomings that occur during photography (lonely points (pixels) that are different from the background, but are not parts images of industrial parts) use checking neighboring pixels for belonging to the background. For this check, we can enter a picture image with values of 0 and 1, where 0 is a point and 1 is not the background [10].

The algorithm for filtering characteristic lines of images of industrial parts is carried out according to the following rules (7):

$$f(i,j) = \begin{cases} 0; \left(\sum_{k=-1}^{1}\sum_{l=-1}^{1} x_{i+k,j+l}\right) < p \\ 1; \left(\sum_{k=-1}^{1}\sum_{l=-1}^{1} x_{i+k,j+l}\right) \geq p \end{cases} \tag{7}$$

$$i = \overline{1, n-1}; j = \overline{1, m-1}; k = l \neq 0; i, j, k, l, p \epsilon Z$$

where f(i,j) is a boolean function that determines whether there are connections to neighboring pixels,

xi,j is mask value for the i-th, j-th pixel;

n is the width of the image in pixels,

m is the height of the image in pixels,

p is the minimum number of adjacent points required to assign a background accessory value to a point.

The number p = 6 was determined experimentally as the most optimal quality/performance criteria.

Moreover, with an increase in the value of the p-parameter, the quality of image processing increases with a decrease in the speed of the program, but with large values of the p parameter, points that may have informative value are discarded. An example of the definition of the parameter p is shown in Fig. 1. For point M(x,y), it is necessary to check adjacent points from point P8 (x−1, y), (since the previous point of the contour was P6), then clockwise P_1(x−1, y+1), P_2(x, y+1), P_3(x+1, y+1), P_4(x+1, y), P_5(x+1, y−1), P_6 (x,y−1), after that the check on the rule (3) can be completed, because p = 6. As can be seen from Fig. 1, the value p = 6 is quite enough to recognize the point as informative.

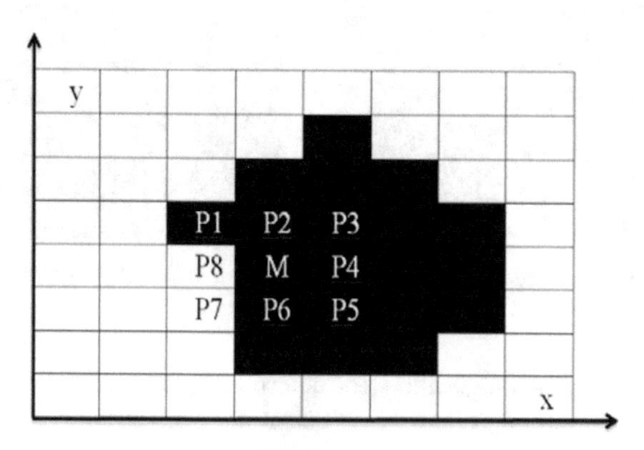

Fig. 1. Determination of p-parameter.

To determine the points of the characteristic lines of the outer boundary (contour) of the images of industrial parts, we can use the following algorithm:

- Choose the starting point;
- Moving clockwise, select the next (neighboring) pixel that does not belong to the background, but borders it;
- Check for the presence of the next (adjacent) point that satisfies the conditions for belonging of the point to the path. If conditions do not satisfy, they go to paragraph 5);
- In case of a positive result of paragraph 3 (the coordinates of the active point have not changed), "rollback" is carried out. The point is assigned a background status, actively selects the previous point and goes to paragraph 2;
- Check at the end of the definition (closure) of the circuit, when the active point returns to its original state using the Jacobs criterion. If the condition is not met, then go to paragraph 2);
- After obtaining a closed contour (the program has successfully selected some integral area of the image other than the background), assign an identifier of industrial details to each point of the image bounded by the contour, in order to avoid re-processing these pixels, as well as to facilitate further work on the image and the selection. After the assignment is complete, go to paragraph 1).

To check the quality of the algorithm, a photograph of the surface of industrial parts was used as a test image (Fig. 2). The result of image processing is shown in Fig. 3.

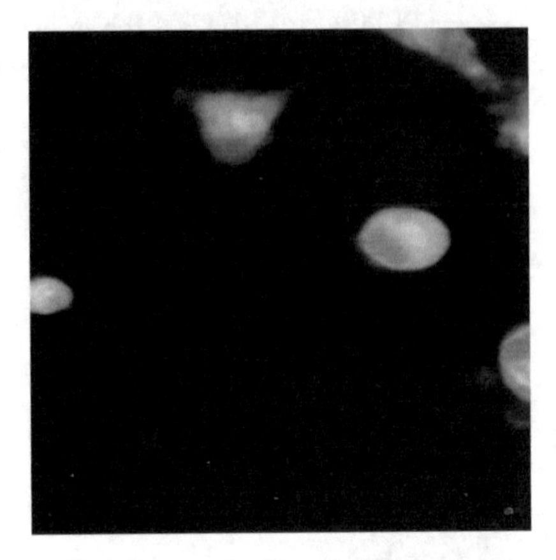

Fig. 2. Photography intended for processing.

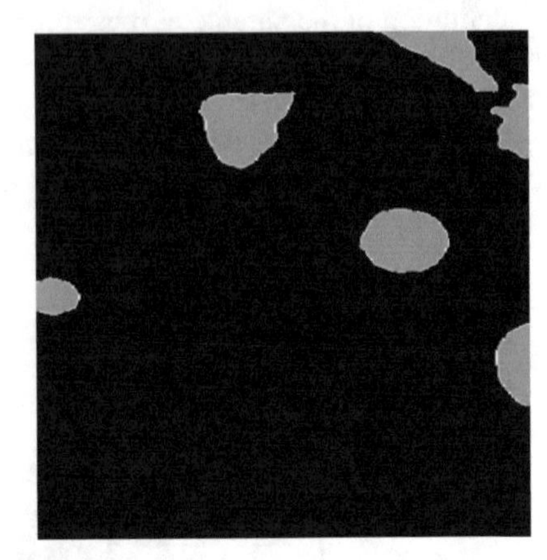

Fig. 3. Processed photo (with selected contours).

Based on the proposed algorithms for determining characteristic lines, a software system in the Delphi 6.0 programming language has been developed [7]. The program identifies the characteristic lines of the external contours of industrial parts' images and conducts a number of studies, including the shape of the surface defects of industrial parts. The defects' shape of the surface of industrial parts is determined by the ratio of the segment connecting the two most distant points of the contour A (x_n, y n) and B (x_{n+m} y_{n+m}) and the perpendicular passing through the middle of the segment AB point M (x_c, y_c), and the point of the contour C (x_g,y_g).

Also, the program, as a result of its work, calculates the geometric parameters of surface defects of industrial parts for further analysis.

3 Conclusions

The paper analyzes the algorithm for selecting characteristic lines for quality control of industrial products and proposes an algorithm for determining the characteristic lines of the external contour (boundary) of industrial parts' images, and also presents the results of the software system. This algorithm can be used to recognize images (characteristic lines of the external contour of the image) in industrial image recognition systems parts. To further develop and increase the efficiency of the process of passing the characteristic lines of the contours of images of industrial parts, it is necessary to develop criteria for the correct shutdown of algorithms, as well as ways to process the protrusions of images (thickness of 1 pixel) more clearly.

References

1. Gavrikov, M.M., Sinetsky, R.M.: Algorithm of one-dimensional structural-approximation analysis of images of cellular surfaces in the problem of input metrological control of industrial products, vol. 6(66), pp. 53–68 (2020)
2. Kuj, S.A., Golovanova, N.B.: On improving the mechanisms for training scientific and pedagogical personnel and the prospects for targeted training in the interests of universities. Russ. Technol. J. **8**(4), 112–128 (2020)
3. Logunova, O.S., Makarychev, P.P.: Algorithms and software for the recognition of low-contrast images in the assessment of steel quality. Softw. Prod. Syst. **3**, 79–81 (2008)
4. Manilo, M.K.: Structural-approximation approach in the problem of photocontrol of cellular surfaces of industrial products. Actual scientific research in the modern world. In: Materials LXXX International Scientific Conference, vol. 12(80), pp. 91–96 (2021)
5. Manilo, M.K.: Methods of structural approximation in various fields of application. In: BSTU Conference. Belgorod, pp. 3673–3677 (2021)
6. Manilo, M.K., Sinetsky, R.M.: Structural-approximation algorithms of preliminary metrological control of cellular surfaces of industrial products. Ser.: Tech. Sci. **3**, 48–56 (2022). University Bulletin North-Caucasian region
7. Ngo, T.-T.: Reflectance and shape estimation with a light field camera under natural illumination. Int. J. Comput. Vision **16**, 1–16 (2020)
8. Nitin, T.: A review: image edge unmasking by applying renovated and colony optimization technique. Int. J. Adv. Res. Comput. Commun. Eng. **4**(6), 35–38 (2015)
9. Sinetsky, R.M., Manilo, M.K.: Application of the method of structural approximation in the problem of photocontrol of cellular surfaces of industrial products. In: Materials of the V National Conference of the Faculty and Researchers of the SRSPU (NPI), vol. 3(6), pp. 37–40 (2020)
10. Urbancic, T.: The influence of the input parameters selection on the RANSAC results. Int. J. Simul. Modell. **13**, 159–170 (2014)

Assessment of the Impact of Technological Trends on Business Activities in the Context of Digital Transformation of Industry

Alla Nikonorova[1], Victoria Perskaya[1], Tao Itao[2], Dmitry Morkovkin[1](✉), Sergey Shmanev[1], and Vagif Kerimov[3]

[1] Financial University Under the Government of the Russian Federation, Moscow, Russia
morkovkinde@mail.ru
[2] Shenzhen University, Shenzhen, Guangdong, China
[3] Moscow State University of Humanities and Economics, Moscow, Russia

Abstract. The prospects for the introduction of digital technologies into the activities of industrial enterprises and the opportunities for the accomplishing the objectives of the sustainable development concept are considered in the article. Particular attention is paid to the study of possibilities of using new technological solutions for improvement of an enterprise's economic performance with simultaneous reducing the negative impact of their activities on the environment. The methods of analogy, synthesis, and extrapolation were used for investigation of technological trends influence in the field of industrial companies' sustainability. The possibilities of usage the advanced experience in introducing digital technologies into manufacturing processes and technological cycles is studied. The results of the analyzes of the modern technological trends influence on the activities of domestic industrial enterprises are described. The model of improvement of industrial enterprises management systems with regard to influence of technological trends and the goals of sustainable development is submitted.

Keywords: Industrial Enterprises · Management · Digital Technologies · Production Processes · Sustainable Development · Technological Trends

1 Introduction

The functioning of industrial enterprises is carried out in difficult conditions of a significant change in the external environment and the emergence of new innovative solutions. The directions of the most intensive development of digital technologies are also changing continuously over time, which increases the importance of studying modern technological trends, tracking and adapting to them by domestic industrial enterprises seeking to maintain their positions in the market, match the level of the most advanced technological developments and preserve the principles of sustainable development. The analysis of the impact of technological trends on the activities of industrial enterprises contributes to successful finding ways and means to strengthen their competitive positions in the market.

© The Author(s), under exclusive license to Springer Nature Switzerland AG 2023
A. Gibadullin (Ed.): DITEM 2022, LNNS 683, pp. 160–168, 2023.
https://doi.org/10.1007/978-3-031-30926-7_16

The aim of the research is to develop a methodology for assessment the technological trends influence on activities of industrial enterprises with regard to achievement of the sustainable development goals. Its development and implementation could be used for determining directions of further development of industrial enterprises. The study of modern technological trends helps to achieve sustainable development goals of industrial enterprises, including measures to reduce their negative impact on the economy.

In modern conditions it is crucial for industrial enterprises to implement the policy of social responsibility in technological progress and strive in achievement of sustainable development goals.

2 Materials and Methods

Tracking trends in technological development provides an opportunity for enterprises to adjust their strategic plans and adopt to changing environmental conditions. The research of the influence of technological trends on activities of industrial enterprises contributes to finding ways and means to strengthen their competitive positions in the market. The scientific works of Mikhailova L.V., Sazonova M.V. [1], Orlov S.N. [2], Chulanova O.L. [3], Kravets A.G., Salnikova N.A. [4] are devoted to researches in this field. The study of technological trends is of interest for not only numerous scientific researchers, but also for well-known consulting companies such as Gartner Incorporated [5–7]. Despite the fact that this activity has been carried out for quite a long period of time, insufficient attention is given to the consideration of the impact of technological trends on the activities of industrial enterprises from the perspective of the sustainable development concept.

The Ministry of Economic Development of the Russian Federation has compiled and published the forecast of socio-economic development of Russia for the planning period of 2022–2023. In accordance with this forecast, the following dynamics are planned in the industrial production of the country (Table 1) [8].

Table 1. Dynamics of industrial production.

% per year	2020	2021	2022	2023	2024	2024/2020
Total volume of industrial production	−2.1	4.2	3.3	2.4	2.2	12.7

According to the assessment of the Ministry of Economic Development, in 2020 the decline in industrial production in the Russian Federation amounted to -2.1%, and in the further years a gradual recovery of industrial production is expected to be observed. Adaptation of global and national economies to the turbulences caused by the pandemic of coronavirus infection redounds to this process.

The study of technological trends and tendencies are in the focus of interests for numerous scientists and researchers. In the process of managing industrial enterprises, it is necessary to take into account modern strategic technological trends. The lists of the most prospectus strategic technological trends have been annually submitted by such well-known consulting company as Gartner Incorporated. The lists of the top strategic technological trends complied by Gartner Incorporated are submitted in the Table 2 [5–7].

Table 2. The top strategic technology trends of Gartner Incorporated.

The strategic technology trends		
2020	2021	2022
1. Hyperautomation 2. Multiexperience 3. Democratization 4. Human augmentation 5. Transparency and traceability 6. The empowered edge 7. The distributed cloud 8. Autonomous things 9. Practical blockchain 10. AI security	1. Internet of behaviors 2. Total experience 3. Privacy-enhancing computation 4. Distributed cloud 5. Anywhere operations 6. Cybersecurity mesh 7. Intelligent composable business 8. AI engineering 9. Hyperautomation	1. Data fabric 2. Cybersecurity mesh 3. Privacy-enhancing computation 4. Cloud-native platforms 5. Composable applications 6. Decision intelligence 7. Hyperautomation 8. AI engineering 9. Distributed enterprises 10. Total experience 11. Autonomic systems 12. Generative AI

The research of the selected theme of technological trends influence on industrial enterprises activities is crucial in conditions of achievement of the Sustainable Development Goals approved and set by the United Nations General Assembly in 2015 for the period up to 2030 [9].

The goals of the present paper are achieved by applying the methods of analogy, synthesis, and extrapolation. The choice of research methods is determined by the specifics of the research theme, as well as the successful approbation of the results of previously conducted researches in the studied area.

The development of digitalization leads to changes in enterprises operating in the real sector of the economy [2]. The active development of innovative technologies creates conditions for successful development of enterprises through the dissemination of Internet technologies, the introduction of electronic information systems and knowledge management tools. It has a significant impact not only on production and management processes, but on the results of industrial enterprises' activities and their influence on environment.

3 Results

Forecasting the development of technological trends becomes the key to the sustainability of an industrial enterprise, it allows minimizing the risks associated with the emergence and penetration of disruptive technologies into production processes [4].

The conducted research was fulfilled within the aim to gain the results suitable for implementation in activities for the sustainable development of industrial enterprises. Their activities may correlate with the goals of sustainable development. Their list and the characteristics of their implementation in practice of industrial enterprises are presented in the Table 3.

Table 3. The Sustainable Development Goals connected with activities of industrial enterprises.

The Goal Number	The Sustainable Development Goal	The form of implementation
Goal 1	No poverty	Providing employment
Goal 3	Good health and well-being	Expanding opportunities to improve the quality of life
Goal 4	Quality education	Development of the competencies necessary for performance of work
Goal 7	Affordable and clean energy	Search and usage of clean energy sources
Goal 8	Decent work and economic growth	Operating modern technical means and increase in efficiency
Goal 9	Industry, innovation and infrastructure	Development and implementation of innovative decisions
Goal 12	Responsible consumption and production	Optimization of production and consumption cycles
Goal 13	Climate action	Implementation and diffusion of climate-friendly technologies
Goal 17	Partnership for the Goals	Development of conditions for successful and versatile partnership

In conditions of introduction of digital technologies, making efforts for compliance to the strategic technological trends, and implementation of modern tools of neural network economy, the manufacturer receives such opportunities as:

- Reduction of operating costs;
- Application of standard models;
- Cost reduction as a result of optimizing the resource supply system;
- Expansion of sales markets and work with individualized demand;
- Reduction of logistics costs due to their optimization;
- Increase in speed of turnover;
- Optimization of management processes and elimination of dual functions;
- Reducing the expenses due to widespread use of remote work technologies.

Schematically, the process of forming an industrial enterprise management system considering influence of strategic technological trends and achievement of the sustainable development goals can be represented as follows (Fig. 1).

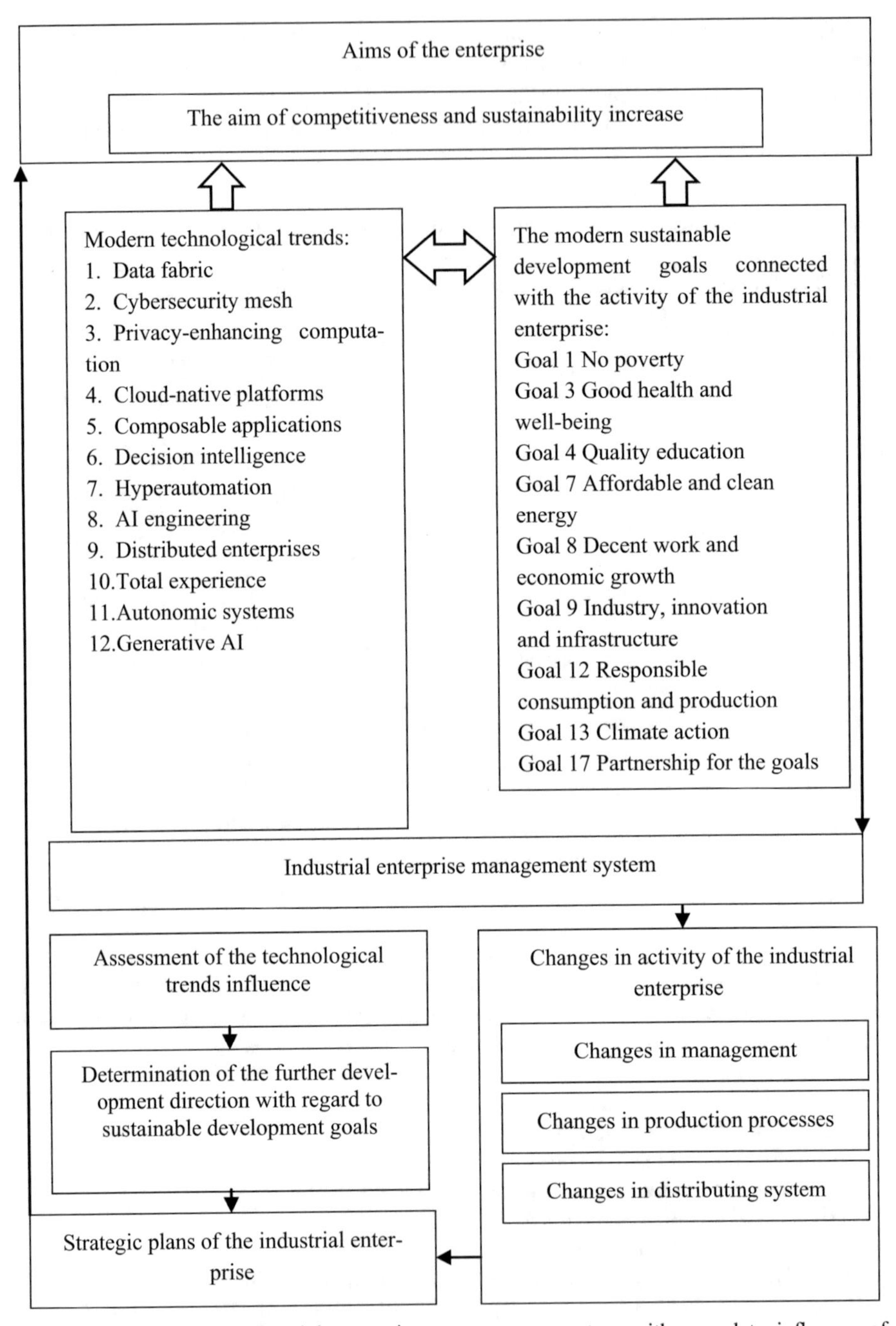

Fig. 1. Functioning of industrial enterprise management system with regard to influence of technological trends and the goals of sustainable development.

The aims of the industrial enterprise are set with regard to opportunities in increase of the enterprise competitiveness owning to implementation of the modern technical decisions and in enforcement of sustainability. The industrial enterprise management system determines and corrects the aims and strategic plans on the basis of assessment of prospects for implementation of the technological trends, the specifics of the industrial enterprise activity, and opportunities in the sustainable development goals achievements.

Economic sustainability and development of industrial enterprises depends on a number of factors. It is possible to distinguish among them:

- Factors that affect the social component of the sustainable development of an industrial enterprise;
- Factors that affect the environmental component of the sustainable development of an enterprise;
- Factors that affect the economic component of the sustainable development of an enterprise.

Among the factors of sustainable economic development of an industrial enterprise it is necessary also to distinguish threats and opportunities of external and internal environment. There are such important indicators as political situation, rational usage of natural resources, production potential of an industrial enterprise, results of market reforms in property relations, improvement of conditions and living environment of population, protection of environmental safety of territory of an enterprise [10].

The analysis of these factors makes it possible to identify the degree of their impact on the growth of economic indicators [11].

Let us designate a number of measures, aimed at increasing the competitiveness of domestic enterprises with regard to the influence of technological trends and the Goals of sustainable development. Among such measures the followings can be highlighted:

- Tracking global trends and correcting the existing strategic plans with regard of them;
- Development and implementation of permanent innovation monitoring systems;
- Planning internal costs for research and development of up-to-date technologies with regard to technological trends and the Goals of sustainable development;
- Periodic and continuous adjustment of strategic plans;
- Development of innovative infrastructure;
- Creating conditions for the development of human capital;
- Environmental protection.

Invention of new technologies helps in simultaneous achievement both of sustainable development goals and compliance to technological trends. As an example of such technology in practice of Russian industrial enterprises it is possible to consider the new method for controlling gas turbine installations to reduce emissions. This method was revealed and submitted by researchers at Perm National Research Polytechnic University (Perm Polytechnic University). The new way to control gas turbine installations helps to reduce carbon monoxide emissions, increase the service life of equipment and the efficiency of its operation [12]. Implementation of this technology is able not only to improve the economic performance of industrial enterprises, but also to reduce the negative impact of their activities on the ecology.

Application of these measures can contribute to increasing the level of stability and sustainability in development of Russian industrial enterprises in conditions of hypercompetitive economy.

4 Discussion

Modernization of the Russian economy and globalization of competition makes it necessary to review of previously selected strategic directions for development and search for new management models. The use of digital technologies and implementation of automating systems makes it possible to use resources of an enterprise more efficiently, to reduce the negative impact on the environment. In the context of sustainable development, the researches of the technological trends influence on activities of industrial enterprises remain crucial and receive new tools to achieve the determined goals. Stabilization of the environmental situation in the region and prevention of industrial enterprises negative impacts are becoming the most efficient in conditions of state support [13]. Formation of digital economy intensifies innovative activity, causes changes in financial and economic mechanisms [14]. Another effective tool for reducing the negative impact of enterprises activities on the environment is the creation of clusters, as it helps to combine the joint efforts of several enterprises in achieving the same goal [15].

In conditions of intensive technological development of foreign countries, the retaining by domestic enterprises of their competitive positions in the market is possible only in case of prompt reaction to the appearance of new technologies. One of the tools for increasing the efficiency of enterprises is the improvement of enterprise management mechanisms with regard to current trends in technology development and modern challenges such as achievement of the goals of sustainable development.

The proposed methodology can be applied in management of Russian industrial enterprises for solving problems caused by modernization in manufacturing and production processes and cycles. It allows contemplating the impact of the external environment on the enterprise, the combination of the influence of technological trends and conditions conducive to the sustainable development of the enterprise, as well as the nature of the impact of the enterprise's activities on the economy. Consideration of these factors contributes to solving numerous problems of industrial enterprises.

The uncertainty of the development of events and the situation in which Russian industrial enterprises operate is a complex problem, the solution of which can be facilitated by a systematic analysis of the ways of technological trends development and the means for achievement of the sustainable development goals.

Production processes and cycles are currently becoming the object for application of digital technologies [1]. In modern conditions the rapid intellectualization of goods and services is observed. Special measures are being taken to minimize the risks of introducing artificial intelligence. Managing the integration of artificial intelligence technologies into work of industrial enterprises allows, in conditions of digital transformation, to make it more adapted to real conditions and understandable for employees involved in production processes [3].

As a result of introduction of information technologies, the direct participation of a person in production and other technical processes decreases, while his involvement in management processes and in areas requiring a creative, informal approach increases.

5 Conclusions

The research of possibilities of technological trends implementation in activities of Russian industrial enterprises and the search for the most promising technological solutions expand the possibilities of adjusting the strategy of their innovative and sustainable development. The processes of introduction of digital technologies into activities of domestic industrial enterprises are reinforced by significant technological progress, behavior of competitors, and changes in the market situation.

All the planned results of the present study have been completely fulfilled. The analysis of the impact of modern technological trends on the activities of Russian industrial enterprises is carried out in the context of sustainable development and with regard to the possibility of reducing their negative impact on the environment. The results of the studies and advanced experience in introduction of digital technologies into production processes and technological cycles of industrial enterprises reveal significant potential for further development.

The list of the technological trends for a particular industrial enterprise depends on its specifics, but the basis for it should be created with regard to the changing over time global tendencies. The activity of industrial enterprises can be at the same time directed to achievement of a complex of the sustainable development goals.

Tracing the aim of creating opportunities for simultaneous solution of several crucial for an industrial enterprise tasks the author's model of improvement of industrial enterprise management system with regard to influence of technological trends and the goals of sustainable development is submitted. It considers the measures, that are capable to be concurrently aimed at maintaining the current rate of digitalization, strengthening the level of sustainability of the company, and reducing negative impact on environment. The results of the present study may be of interest for senior and middle level management of industrial enterprises, for manager involved in the process of development and improvement of the development strategy of a manufacturing enterprise. Achievement of increase in economic indicators of an enterprise with a simultaneous reduction in negative impact of its activity on environment is possible only in case of purposeful work of management of an enterprise, rapid identification, and introduction of modern innovative technologies that contribute to rationalization and optimization of production processes.

References

1. Mikhaylova, L.V.б., Sazonova, M.V.: Digitalization of production in the Russian economy: theoretical analysis and development trends. Vestnik Moskovskogo gosudarstvennogo oblastnogo universiteta. Seriya: Ekonomika **4**, 57–63 (2021)
2. Orlov, S.N.: Organizational and technological trends of digital economy. Finansovaya Ekonomika **10**, 437–439 (2019)

3. Chulanova, O.L., Khaybullova, K.N.: Managing integration of artificial intelligence technologies as a technological trend in the context of digital transformation into work with personnel. Vestnik Surgutskogo Gosudarstvennogo Universiteta **1**(27), 112–121 (2020)
4. Kravets, A.G., Salnikova, N.A.: Predictive modeling of trends in technological development. Izvestiya Sankt-Peterburgskogo Gosudarstvennogo Tekhnologicheskogo Instituta (Tekhnicheskogo Universiteta) **55**, 103–108 (2020)
5. Gartner Top 10 Strategic Technology Trends for 2020. https://www.gartner.com/smarterwithgartner/gartner-top-10-strategic-technology-trends-for-2020. Accessed 29 Jan 2022
6. Gartner's Top Nine Strategic Tech Trends For 2021. https://www.forbes.com/sites/peterhigh/2020/10/26/gartners-top-nine-strategic-tech-trends-for-2021/?sh=5d7a67e821f6. Accessed 29 Jan 2022
7. Gartner Top Strategic Technology Trends for 2022. https://www.gartner.com/en/information-technology/insights/top-technology-trends. Accessed 29 Jan 2022
8. Ministry of Economic Development of the Russian Federation. Forecasts of socio-economic development. https://www.economy.gov.ru/material/directions/makroec/prognozy_socialno_ekonomicheskogo_razvitiya/prognoz_socialno_ekonomicheskogo_razvitiya_rf_na_2022_god_i_na_planovyy_period_2023_i_2024_godov.html. Accessed 01 Feb 2022
9. Chugumbayev, R.R., Chadayeva, A.V.: Influence of sustainable development goals on the formation of a business development strategy. Ekonomika. Sotsiologiya. Pravo **2**(22), 48–55 (2021)
10. Koryakov, A.G.: Factors of economic stability and development of enterprises. Nalogi i nalogooblozheniye **5**, 55–61 (2012)
11. Fedorova, S.N., Razzhivin, O.A., Zamkovoy, A.A., Potapova, E.V.: Characteristic of economic indicators of reproduction of fixed capital. Int. J. Appl. Bus. Econ. Res. **12**, 73–82 (2017)
12. Perepelitsa, E.: Russian chemists have developed a new method of carbon dioxide capture. https://inscience.news/ru/article/russian-science/8340. Accessed 07 Feb 2022
13. Veselovskiy, M.Y.: Improving state support for domestic corporations. Voprosy Regionalnoy Ekonomiki **2**, 78–82 (2012)
14. Veselovsky, M.Y., Pogodina, T.V., Ilyukhina, R.V., Sigunova, T.A., Kuzovleva, N.F.: Financial and economic mechanisms of promoting innovative activity in the context of the digital economy formation. Entrepreneurship Sustain. **5**(3), 672–681 (2018)
15. Kiseleva, N.V., Panichkina, M.V., Klochko, E.N., Nikonorova, A.V., Kireev, S.V.: Creation of clusters of small enterprises of the region. Int. J. Econ. Financ. **6**(2), 294–297 (2016)
16. Zakirova, A., Klychova, G., Ostaev, G., Zalilova, Z., Klychova, A.: Analytical support of management accounting in managing sustainable development of agricultural organizations. In: E3S Web of Conferences, vol. 164, p. 10008 (2020)
17. Lukyanova, M.T., Kovshov, V.A., Galin, Z.A., Zalilova, Z.A., Stovba, E.V.: Scenario method of strategic planning and forecasting the development of the rural economy in agricultural complex. Scientifica **2020**, 9124641 (2020)
18. Orazbayev, B., Ospanov, E., Kissikova, N., Mukataev, N., Orazbayeva, K.: Decision-making in the fuzzy environment on the basis of various compromise schemes. Procedia Comput. Sci. **120**, 945–952 (2017)
19. Romanova, J.A., Morkovkin, D.E., Romanova, I.N., Artamonova, K.A., Gibadullin, A.A.: Formation of a digital agricultural development system. In: IOP Conference Series: Earth and Environmental Science, vol. 548, p. 032014 (2020)

Author Index

© The Editor(s) (if applicable) and The Author(s), under exclusive license
to Springer Nature Switzerland AG 2023

A. Gibadullin (Ed.): DITEM 2022, LNNS 683, pp. 169–170, 2023.
https://doi.org/10.1007/978-3-031-30926-7

Printed in the United States
by Baker & Taylor Publisher Services